Lecture Notes in Mathematics 1679

Springer
Berlin
Heidelberg
New York
Barcelona
Budapest
Hong Kong
London
Milan
Paris
Santa Clara
Singapore
Tokyo

Karl-Goswin Grosse-Erdmann

The Blocking Technique, Weighted Mean Operators and Hardy's Inequality

Springer

Author

Karl-Goswin Grosse-Erdmann
Fachbereich Mathematik
Fernuniversität Hagen
Postfach 940
D-58084 Hagen, Germany
e-mail: kg.grosse-erdmann@fernuni-hagen.de

Cataloging-in-Publication Data applied for

Die Deutsche Bibliothek - CIP-Einheitsaufnahme

Grosse-Erdmann, Karl-Goswin:
The blocking technique : weighted mean operators and Hardy's inequality / Karl-Goswin Grosse-Erdmann. - Berlin ; Heidelberg ; New York ; Barcelona ; Budapest ; Hong Kong ; London ; Milan ; Paris ; Santa Clara ; Singapore ; Tokyo : Springer, 1998
(Lecture notes in mathematics ; 1679)
ISBN 3-540-63902-0

Mathematics Subject Classification (1991):
26D15, 40A05, 40A10, 40G05, 46B45, 46E30, 47B37

ISSN 0075-8434
ISBN 3-540-63902-0 Springer-Verlag Berlin Heidelberg New York

Typesetting: Camera-ready TEX output by the author
SPIN: 10553437 46/3143-543210 - Printed on acid-free paper

Für Klaus

Preface

The aim of these notes is to present a comprehensive treatment of the so-called blocking technique, together with applications to the study of sequence and function spaces, to the study of operators between such spaces, and to classical inequalities.

In these theories, and in other parts of Analysis, expressions of the form

$$\sum_{n=1}^{\infty}\left[a_n\left(\sum_{k=1}^{n}|x_k|^p\right)^{1/p}\right]^q$$

play an important rôle, most prominently perhaps in connection with Hardy's inequality. The analysis of such an expression, which we shall briefly call a *norm in section form*, has turned out to be demanding.

In many cases a problem becomes more accessible under a suitable renorming. Now, throughout the last four decades expressions of the form

$$\sum_{\nu=0}^{\infty}\left[\frac{1}{2^{\nu\alpha}}\left(\sum_{k\in I_\nu}|x_k|^p\right)^{1/p}\right]^q$$

have been appearing quite naturally in various parts of Analysis, very often in connection with coefficient conditions on series expansions of functions. Here, the I_ν form a partition of $\mathbb{N}$ into disjoint intervals, the most common partition being that into the dyadic blocks $[2^\nu, 2^{\nu+1})$. An expression of the above type is called a *norm in block form*.

It has already been noted by several authors that certain norms in section form can be replaced equivalently by a norm in block form. Such a renorming, which is referred to as the *blocking technique*, is of great practical value, for the analysis of norms in block form is much simpler: in many respects they behave just like the familiar l^p-norms.

In these notes we show that, apart from some trivial cases, in fact every norm in section form can be transformed into block form and, what is perhaps even more surprising, every norm in block form can be re-translated into section form. In that sense the blocking technique is universal. Chapter I provides a dictionary of transformations between the two kinds of norms. The related problem of characterising when two given norms are equivalent is of less relevance to the applications in these notes and is treated in the Appendix. In Chapter II we apply the blocking technique to study the structure of sequence spaces defined by norms in section form, while Chapter III contains applications to (generalised) weighted mean operators in l^p and to the weighted inequalities of Hardy and Copson.

It is more a matter of personal taste that we have chosen to concentrate our study on norms for sequences rather than on the corresponding integral norms for functions on the real line. In Chapter IV we indicate the integral analogues of our results.

Our research originated from a study of four papers by G. Bennett that revolve around the inequalities of Hardy and Copson. We have developed the blocking technique as a tool to attack some of his open problems. This has been successful; the solutions to three of his problems are contained in Sections 9, 10 and 17.

On the other hand, the results in Bennett's papers were instrumental in leading us to the appropriate transformations between section norms and block norms. Thus it is in two ways that these notes owe their existence to Grahame Bennett. I would therefore like to take this opportunity to express my sincere gratitude to him and my deep appreciation of the beauty of his work.

Hagen, October 1997 Karl-Goswin Grosse-Erdmann

Contents

Introduction

In four fundamental papers G. Bennett [12, 13, 14, 15] has undertaken a thorough investigation of the inequalities of Hardy and Copson and their weighted generalisations. Among other things he has completely solved the l^p-mapping problem for weighted mean matrices and, in [15], has introduced the new concept of factorisation of inequalities. This concept, for instance, allows Hardy's classical inequality to be seen in a new light, 75 years after its first appearance. Bennett has also formulated various open problems that were raised by his work. A new approach, the so-called blocking technique, has enabled us to solve three of his problems, and it turned out that this technique also serves to obtain a large part of Bennett's results in an elementary and unified way, as far as its *qualitative* aspect is concerned (we shall say more about this point below).

Since our investigation revolves around Bennett's four papers we shall refer to them throughout briefly as **BI, BII, BIII** and **BIV**. By means of Hardy's inequality we shall next illustrate what we mean by the blocking technique and how it comes into play.

Hardy's inequality

This inequality, in its discrete form, asserts that for any $p > 1$ there is some constant $K > 0$ such that

$$\sum_{n=1}^{N} \left(\frac{1}{n} \sum_{k=1}^{n} x_k \right)^p \leq K \sum_{n=1}^{N} x_n^p \tag{0.1}$$

holds for every $N \in \mathbb{N}$ and all non-negative numbers $x_1, \ldots, x_N$. Letting $N \to \infty$ we see that this immediately implies the inclusion

$$l^p \subset \operatorname{ces}(p) \tag{0.2}$$

between the space l^p of p-summable (real or complex) sequences and the so-called *Cesàro sequence space*

$$\operatorname{ces}(p) = \left\{ x = (x_k) : \sum_n \left(\frac{1}{n} \sum_{k=1}^{n} |x_k| \right)^p < \infty \right\}.$$

As a matter of fact, the inclusion (0.2) is the form in which Hardy [36] first announced his result. Since $\operatorname{ces}(p)$ is a Banach space under the norm

$$\|x\|_{\operatorname{ces}(p)} = \left(\sum_n \left(\frac{1}{n} \sum_{k=1}^{n} |x_k| \right)^p \right)^{1/p}, \tag{0.3}$$

an application of the closed graph theorem shows that (0.2) implies (0.1) so that the two are in fact equivalent.

The norm of $\operatorname{ces}(p)$, although at first sight a rather straightforward variation of the l^p-norm, defies simple analysis. This is seen most clearly in a result of A. A. Jagers [44]. Answering a *prijsvraag* of the Dutch Mathematical Society [101], Jagers determined the dual of $\operatorname{ces}(p)$ under its dual norm; it turned out to be more complicated than one would expect. One of Bennett's numerous surprising results is that in fact

$$\operatorname{ces}(p)^* \cong \left\{ x : \sum_n \sup_{k \geq n} |x_k|^{p^*} < \infty \right\}, \tag{0.4}$$

where p^* is the conjugate exponent to p (**BIV**, 12.17). The price one has to pay for this simple representation is that the norm on $\operatorname{ces}(p)^*$ implied by it is *not* the dual norm. Bennett rests the proof of (0.4) on a renorming of the Cesàro sequence space, the new norm being suggested by a factorisation result.

The main idea of these notes is to renorm $\operatorname{ces}(p)$ in another direction. What makes the analysis under (0.3) difficult is the fact that each term of the sequence x appears in almost every expression $\frac{1}{n} \sum_{k=1}^{n} |x_k|$. It would considerably simplify matters if this norm could be replaced by one in which each x_k only appeared once. This is indeed possible; we shall show that an equivalent norm on $\operatorname{ces}(p)$ is given by

$$\|x\| = \left(\sum_{\nu=0}^{\infty} 2^{\nu(1-p)} \left(\sum_{k=2^\nu}^{2^{\nu+1}-1} |x_k| \right)^p \right)^{1/p}, \tag{0.5}$$

and that ces(p) consists of all sequences $\boldsymbol{x}$ for which this new (extended) norm is finite, see Theorem 4.1. Since we can write

$$\|\boldsymbol{x}\|_{l^p} = \left(\sum_{\nu=0}^{\infty}\Big(\sum_{k=2^\nu}^{2^{\nu+1}-1} |x_k|^p\Big)^{p/p}\right)^{1/p},$$

a simple application of Hölder's inequality shows that (0.2) holds. We have thus obtained a new proof of Hardy's inequality.

The spaces $l(p, q)$

We shall say that a norm like (0.5) is in *block form*, while (0.3) is in *section form.* In 1969, Hedlund [41] introduced the mixed norm spaces

$$l(p,q) = \left\{\boldsymbol{x} : \sum_{\nu=0}^{\infty}\Big(\sum_{k=2^\nu}^{2^{\nu+1}-1} |x_k|^p\Big)^{q/p} < \infty\right\},$$

see also Kellogg [51]. In many respects these spaces behave just like the familiar l^p-spaces. What we have found is that the Cesàro sequence space ces(p) is a weighted $l(1,p)$-space. This also helps to locate the place of ces(p) within the collection of classical and semi-classical Banach sequence spaces (cf. **BIV**, p. 7). We remark that the Besov sequence spaces $b^{\rho}_{p,q}$ introduced by Pietsch [82] in 1980 are weighted $l(p,q)$-spaces with, in particular, $b^0_{p,q} = l(p,q)$.

The blocking technique: scope and limitations

The *blocking technique* consists in replacing norms in section form by norms in block form and vice versa. In our applications of this technique the rôle of the norms in block form is that of a catalyst. We start off with problems that are formulated in terms of section norms, translate these into block form, solve the new and usually much simpler problems, and re-translate the solution into section form. Thus, for example, questions on spaces like ces(p) are reduced to questions on $l(p,q)$-spaces. The main difficulty in this programme consists in finding a suitably large number of transformations between section and block form that is flexible enough for differing purposes.

We shall see that our approach not only provides a new proof of Hardy's inequality (and of the related inequality of Copson, compare Section 10), but that its scope is much wider. Among other things it enables us to treat these inequalities in their weighted form. For this additional generality, however, one has to depart from the dyadic blocks $[2^\nu, 2^{\nu+1})$ and has to allow general blocks

$[m_\nu, m_{\nu+1})$. In addition there is an analogue of the blocking technique for integral norms of functions on the real line, so that everything that can be done for series can also be done for integrals.

In particular we shall here prove a conjecture of Bennett (**BIII**, pp. 160-161) and answer two of his open problems (**BII**, p. 393; **BIV**, p. 37). In addition, the blocking technique leads to elementary and unified proofs for a large part of Bennett's investigations in his four papers as far as its qualitative aspect is concerned. This is especially interesting at those points where Bennett uses deep functional analytic techniques, for example in **BIII**.

Another aspect of our work is more "philosophical". A striking feature of Bennett's papers is that many of the problems considered by him have surprisingly simple answers, for example, his solution of the ces(p)-duality problem stated above; Bennett himself expresses his surprise at various places (**BIV**, pp. 2, 26, 68; see also [87, Introduction]). The present notes offer an explanation for this phenomenon: The spaces we are dealing with are, in reality, $l(p,q)$-spaces in disguise, and these spaces are rather well-behaved.

The blocking technique, however, also has its limitations, and it is important to be clear about this. By renorming the spaces involved we lose control over constants, for instance in inequalities, while the major and deeper part of Bennett's work is devoted to finding best-possible constants. Thus, for example, we can give a new proof of Hardy's inequality in its qualitative form (0.1), but we are not able to confirm Landau's result [55] that K can be taken as $(\frac{p}{p-1})^p$, which is best-possible. In that respect our work is merely qualitative. And each of our results in turn poses a new problem: that of finding the hidden best-possible constants.

The blocking technique in the literature

Norms in block form have been appearing in the literature for some time, and with it the blocking technique. The phrase "blocking technique" (or rather blocking method) was suggested by L. Leindler in a recent publication [63, Abstract] in a related context. It is also Leindler who has contributed a large number of equivalence results between section norms and block norms over the past decades, see [59, 65] and the literature cited therein.

Closest in spirit to our work is the use of the blocking technique in connection with spaces of strongly Cesàro summable sequences. These investigations, which were started by Taberski [92], Borwein [21] and Kuttner and Maddox [54] in the early 60's, were our main source of inspiration. For a recent survey see [71].

Norms in block form and the blocking technique also seem to come up naturally when the coefficients in series expansions of functions are studied. They play a major rôle, for instance, in connection with absolute summability of orthogonal series (first appearance in Tandori [96], 1960; for a recent contribution see [65]), multipliers between spaces of analytic functions (Hedlund [41], 1969; recently [17] and [3]), spaces of Fourier coefficients of L^1-functions (Fomin [32], 1978; recently [25] and [6]) and power series with positive coefficients (Mateljević and Pavlović [74], 1983; recently [64]). A very recent addition is the theory of wavelets (see, for example, [75, 6.10]). Thus in all of these areas our results are of relevance.

The present text is the first to apply the blocking technique systematically in the context of Hardy's inequality. The usefulness of the technique in this area was first observed in **BIV**, p. 81, but we shall see that we have to go beyond dyadic blocks in general. We also offer the first comprehensive treatment of the blocking technique itself. There is closely related work due to Totik and Vincze [97] and Leindler [65]. They characterise when two *given* norms, one in section form and one in block form, are equivalent (see also Section 4 and the Appendix). The problem comes to life again, however, if one is given a norm in one of the two forms and has to find an equivalent one in the other form, the transformation from block form into section form posing the main difficulty. This accounts for the fact that some authors have presented their results in block form and have failed to re-translate them into the more natural section form. These re-translations are not at all obvious. They were suggested to us by Bennett's various results.

There are analogues of the norms in block form for functions on the real line (or, more generally, on $\mathbb{R}^n$). Such norms appear abundantly in Harmonic Analysis where they have led to the notion of amalgams. We refer to [33] for a thorough survey and also to the work of Feichtinger, see, for example, [30, 31]. Instances of the blocking technique for functions can be found in connection with the notion of a Lebesgue point (Tandori [94]), in the context of strong Cesàro summability of functions (Borwein [21]) and in the theory of Beurling algebras and more general function spaces (Gilbert [34], Johnson [47]), among others.

Contents

These notes are divided into four chapters and an appendix.

Chapter I develops the blocking technique as indicated above. In Section 2 we obtain transformations from block form into section form while in Section 3 we go the opposite direction. This forms the basis of all that follows.

In Chapter II we introduce two classes of sequence spaces, $c(\boldsymbol{a}, p, q)$ and $d(\boldsymbol{a}, p, q)$. These spaces contain as special cases the space $\mathrm{ces}(p)$ and many other spaces. We apply the results of Chapter I to study their basic structure (Section

6), to characterise the multipliers between these spaces and l^p (Section 7) and to obtain some factorisation results (Section 8).

Chapter III presents our main applications. In Section 9 we deal with Hardy's inequality with weights or, in a different language, with factorable matrices as operators on l^p; these cover the weighted mean matrices as most important special case. In particular we prove the conjecture of Bennett mentioned above. Section 10 studies Copson's inequality with weights, where we also answer another problem of Bennett. In Sections 11 and 12 we treat the reverses of some classical inequalities. We end the chapter with a selection of further applications (Section 13).

In Chapter IV we indicate integral analogues of our results. In particular we complete an investigation started by Beesack and Heinig [9], thus answering a third question of Bennett (**BIV**, p. 37).

We end these notes with an Appendix in which we apply the results of Chapter I to study the equivalence of norms in section form with norms in block form à la Totik-Vincze and Leindler.

Notation

We agree that Roman indices $n, k, \ldots$ start from 1 while Greek indices $\nu, \mu, \ldots$ start from 0, if nothing else is said.

For any sequence $\boldsymbol{x} = (x_n)$ we denote by $P_n\boldsymbol{x} = (x_1, \ldots, x_n, 0, 0, \ldots)$ its nth section.

The space ω is the space of all (real or complex) sequences, its subspace φ consists of all finite sequences. If E is a sequence space, any non-negative sequence $\boldsymbol{w}$ defines a weighted space $E_{\boldsymbol{w}} = \{\boldsymbol{x} : \boldsymbol{x} \cdot \boldsymbol{w} \in E\}$, where the product of two sequences is taken coordinatewise.

For $0 < p \leq \infty$ we define its *conjugate* p^* by $\frac{1}{p} + \frac{1}{p^*} = 1$, with the usual convention if $p = 1$ or $p = \infty$. We remark that $p^* < 0$ if $p < 1$.

As usual the constant K appearing in inequalities may vary from occurrence to occurrence.

Further notation will be introduced at the beginnings of Sections 1, 5, 7 and 8. In addition, we shall adopt the following convention: When a condition contains a sum $\sum_{k=n}^{\infty} c_k$ over non-negative numbers, then the condition is understood to imply that $\sum_k c_k < \infty$. The same applies to a supremum. Further conventions will be introduced in Remark 2.2(i).

Chapter I

The Blocking Technique

1 Norms in Section Form and Norms in Block Form

As we have seen, the norm

$$\|\boldsymbol{x}\|_{\mathrm{ces}(p)} = \left(\sum_n \left(\frac{1}{n}\sum_{k=1}^n |x_k|\right)^p\right)^{1/p}$$

is at the heart of Hardy's inequality. Generalising this we assume that $\boldsymbol{a} = (a_n)$ is any sequence of non-negative terms and that $0 < p, q \leq \infty$. Then, for any sequence $\boldsymbol{x} = (x_n)$, we consider

$$\|\boldsymbol{x}\| = \left(\sum_n \left[a_n \left(\sum_{k=1}^n |x_k|^p\right)^{1/p}\right]^q\right)^{1/q}$$

and its companion

$$\|\boldsymbol{x}\| = \left(\sum_n \left[a_n \left(\sum_{k=n}^\infty |x_k|^p\right)^{1/p}\right]^q\right)^{1/q},$$

with the usual modifications if p or q is infinite. By abuse of language we shall refer to these extended quasi-seminorms briefly as norms, and we say that these norms are in *section form*. At times we shall allow additional weights, that is, we replace x_k by $w_k x_k$.

Our aim is to transform these norms into block form. It turns out that it does not suffice to consider dyadic blocks only. Thus let $\boldsymbol{m} = (m_\nu)_{\nu \geq 0}$ be any *index sequence*, that is, any sequence of integers with $m_0 = 1$, $m_{\nu+1} \geq m_\nu$ for all ν and $m_\nu \to \infty$ as $\nu \to \infty$. The *blocks* associated with $\boldsymbol{m}$ are defined as

$$I_\nu = [m_\nu, m_{\nu+1}) = \{n \in \mathbb{N} : m_\nu \leq n < m_{\nu+1}\}.$$

We allow the I_ν to be empty. The commonest blocks to be found in the literature are the *dyadic blocks* defined by $m_\nu = 2^\nu$. Now let $\alpha \in \mathbb{R}$ and $0 < p, q \leq \infty$. Then, for any sequence $\boldsymbol{x} = (x_n)$, we consider

$$\|\boldsymbol{x}\| = \left(\sum_\nu \left[\frac{1}{2^{\nu\alpha}}\left(\sum_{k\in I_\nu} |x_k|^p\right)^{1/p}\right]^q\right)^{1/q},$$

again with modifications if p or q is infinite. We call this a norm in *block form*.

We start by transforming norms from block form into section form because this turns out to be the more difficult direction and it immediately implies the opposite direction. We postpone a discussion of these results to Section 4.

2 Transformation from Block Form into Section Form

We first define a correlation between index sequences $\boldsymbol{m} = (m_\nu)$ and positive monotonic sequences $\boldsymbol{s} = (s_n)$ with $s_n \to 0$. We say that $\boldsymbol{m}$ and $\boldsymbol{s}$ are *correlated* if

$$(2.1) \qquad \begin{aligned} \frac{1}{2^\nu} \geq s_n &> \frac{1}{2^{\nu+1}} && \text{if } m_\nu \leq n < m_{\nu+1} \quad (\nu \geq 1), \\ s_n &> \frac{1}{2} && \text{if } n < m_1. \end{aligned}$$

Given any such sequence $\boldsymbol{s}$ we see that there is a unique index sequence $\boldsymbol{m}$ correlated to it; it is defined by $m_0 = 1$ and, for $\nu \geq 1$,

$$(2.2) \qquad m_\nu = \min\left\{n : s_n \leq \frac{1}{2^\nu}\right\}.$$

Conversely, to any index sequence $\boldsymbol{m}$ there are infinitely many sequences $\boldsymbol{s}$ correlated to it, for instance the one defined by

$$s_n = \frac{1}{2^\nu} \quad \text{for} \quad m_\nu \leq n < m_{\nu+1}, \nu \geq 0.$$

Throughout this section, let $\boldsymbol{m} = (m_\nu)$ be a fixed index sequence and $\boldsymbol{s} = (s_n)$ a fixed positive monotonic sequence converging to 0 that is correlated to $\boldsymbol{m}$.

Theorem 2.1 *Let* $0 < p, q < \infty$ *and* $\alpha \in \mathbb{R}$. *Then, for any sequence* $\boldsymbol{x} = (x_n)$, *the condition*

$$\sum_\nu \left[\frac{1}{2^{\nu\alpha}}\left(\sum_{k\in I_\nu}|x_k|^p\right)^{1/p}\right]^q < \infty \tag{2.3}$$

is equivalent to any of the following conditions, where $\beta \neq 0, \gamma \neq 0$ *and* δ *are real numbers with* $\gamma/q + \delta/p = \alpha$:

$$\text{(2.4i)} \qquad \sum_n (s_n^\beta - s_{n+1}^\beta)s_n^{\gamma-\beta}\left(\sum_{k=1}^n s_k^\delta |x_k|^p\right)^{q/p} < \infty \qquad \text{if } \beta > 0, \gamma > 0,$$

$$\text{(2.4ii)} \qquad \sum_n (s_{n+1}^\beta - s_n^\beta)s_n^\gamma s_{n+1}^{-\beta}\left(\sum_{k=1}^n s_k^\delta |x_k|^p\right)^{q/p} < \infty \qquad \text{if } \beta < 0, \gamma > 0,$$

$$\text{(2.4iii)} \qquad \sum_n (s_n^\beta - s_{n-1}^\beta)s_n^{\gamma-\beta}\left(\sum_{k=n}^\infty s_k^\delta |x_k|^p\right)^{q/p} < \infty \qquad \text{if } \beta < 0, \gamma < 0,$$

$$\text{(2.4iv)} \qquad \sum_n (s_{n-1}^\beta - s_n^\beta)s_n^\gamma s_{n-1}^{-\beta}\left(\sum_{k=n}^\infty s_k^\delta |x_k|^p\right)^{q/p} < \infty \qquad \text{if } \beta > 0, \gamma < 0.$$

In addition, for real numbers $\gamma \neq 0$ *and* δ *with* $\gamma/q+\delta/p = \alpha$, (2.3) *is equivalent to*

$$\text{(2.4v)} \qquad \sum_n s_n^{\gamma+\delta}|x_n|^p\left(\sum_{k=1}^n s_k^\delta |x_k|^p\right)^{q/p-1} < \infty \qquad \text{if } \gamma > 0,$$

$$\text{(2.4vi)} \qquad \sum_n s_n^{\gamma+\delta}|x_n|^p\left(\sum_{k=n}^\infty s_k^\delta |x_k|^p\right)^{q/p-1} < \infty \qquad \text{if } \gamma < 0.$$

Remark 2.2 (i) In conditions (2.4iii) and (2.4iv) the coefficient for $n = 1$ is undefined. Obviously, its value has no influence on the validity of the theorem. However, it turns out that the most natural choice is to take its value as s_1^γ, that is, to choose $s_0 = \infty$, if one likes. In later sections, s_n will be substituted by certain expressions in a_k. As a consequence, we shall there interpret $\sum_{k=0}^\infty a_k^q$ as ∞ and, as usual, $\sum_{k=1}^0 a_k^q$ as 0, similarly for suprema. In conditions (2.4v) and (2.4vi), $0 \cdot 0^r$ has to be interpreted as 0 even if $r < 0$. *These interpretations are in effect throughout these notes.*

(ii) If one is given an abstract sequence (m_ν) (or (s_n)) and if $\alpha \neq 0$, then the simplest(-looking) equivalent condition for (2.3) is obtained by taking $\beta = \gamma = \alpha q$ and $\delta = 0$:

$$\text{(2.5i)} \qquad \sum_n (s_n^{\alpha q} - s_{n+1}^{\alpha q}) \left(\sum_{k=1}^{n} |x_k|^p \right)^{q/p} < \infty \qquad \text{if} \quad \alpha > 0,$$

$$\text{(2.5ii)} \qquad \sum_n (s_n^{\alpha q} - s_{n-1}^{\alpha q}) \left(\sum_{k=n}^{\infty} |x_k|^p \right)^{q/p} < \infty \qquad \text{if} \quad \alpha < 0.$$

However, as we shall see in Chapter II, in many applications one is confronted with a specific sequence (m_ν) (or (s_n)), in which case it will be useful to have available the additional parameters β, γ and δ.

Proof of Theorem 2.1. Replacing $|x_k|^p$ by $|x_k|$, q/p by q and αp by α shows that we need only consider the case $p = 1$.

(2.3)$\Rightarrow$(2.4i). First let $q > 1$. Then, by (2.1), we have $1/2^{\nu+1} < s_n \leq 1/2^\nu$ for $n \in I_\nu$ and $\nu \geq 1$, hence

$$\begin{aligned} &\sum_{n \in I_\nu} (s_n^\beta - s_{n+1}^\beta) s_n^{\gamma - \beta} \left(\sum_{k=1}^{n} s_k^\delta |x_k| \right)^q \\ &\qquad \leq K \sum_{n \in I_\nu} (s_n^\beta - s_{n+1}^\beta) \frac{1}{2^{\nu(\gamma-\beta)}} \left(\sum_{k=1}^{m_{\nu+1}-1} s_k^\delta |x_k| \right)^q \\ &\qquad \leq K (s_{m_\nu}^\beta - s_{m_{\nu+1}}^\beta) \frac{1}{2^{\nu(\gamma-\beta)}} \left(\sum_{\mu=0}^{\nu} \sum_{k \in I_\mu} \frac{1}{2^{\mu\delta}} |x_k| \right)^q, \end{aligned}$$

now using $\delta = \alpha - \gamma/q$,

$$\leq K \frac{1}{2^{\nu\gamma}} \left(\sum_{\mu=0}^{\nu} \frac{2^{\mu\gamma/q}}{2^{\nu\gamma/q}} \frac{2^{\nu\gamma/q}}{2^{\mu\alpha}} \sum_{k \in I_\mu} |x_k| \right)^q;$$

since $\gamma > 0$, we have $\sum_{\mu=0}^{\nu} 2^{\mu\gamma/q} \sim 2^{\nu\gamma/q}$, so that we can continue by applying Jensen's inequality:

$$\leq K \frac{1}{2^{\nu\gamma}} \sum_{\mu=0}^{\nu} \frac{2^{\mu\gamma/q}}{2^{\nu\gamma/q}} \left(\frac{2^{\nu\gamma/q}}{2^{\mu\alpha}} \sum_{k \in I_\mu} |x_k| \right)^q.$$

The constants K are independent of ν, so that summing on ν gives

$$\sum_{n=m_1}^{\infty}(s_n^\beta - s_{n+1}^\beta)s_n^{\gamma-\beta}\left(\sum_{k=1}^{n} s_k^\delta |x_k|\right)^q \leq K\sum_{\nu=1}^{\infty}\frac{1}{2^{\nu\gamma/q}}\sum_{\mu=0}^{\nu}2^{\mu\gamma/q}\left(\frac{1}{2^{\mu\alpha}}\sum_{k\in I_\mu}|x_k|\right)^q$$
$$\leq K\sum_{\mu=0}^{\infty}\left(\sum_{\nu=\mu}^{\infty}\frac{2^{\mu\gamma/q}}{2^{\nu\gamma/q}}\right)\left(\frac{1}{2^{\mu\alpha}}\sum_{k\in I_\mu}|x_k|\right)^q$$
$$\leq K\sum_{\mu=0}^{\infty}\left(\frac{1}{2^{\mu\alpha}}\sum_{k\in I_\mu}|x_k|\right)^q .$$

The same proof works for $q \leq 1$ if we replace Jensen's inequality by the inequality $(\sum_k c_k)^q \leq \sum_k c_k^q$ for non-negative c_k.

(2.4i)$\Rightarrow$(2.3). Fix $\nu \geq 1$ and assume that $I_\nu \neq \emptyset$. Then $m_{\nu+1} - 1 \neq 0$, so that we deduce from (2.1) that $s_{m_{\nu+1}-1}^\beta - s_{m_{\nu+2}}^\beta \geq K/2^{\nu\beta}$, and hence

$$\left(\frac{1}{2^{\nu\alpha}}\sum_{k\in I_\nu}|x_k|\right)^q = \frac{1}{2^{\nu\beta}}\frac{1}{2^{\nu(\gamma-\beta)}}\left(\frac{1}{2^{\nu\delta}}\sum_{k\in I_\nu}|x_k|\right)^q$$
$$\leq K\left(s_{m_{\nu+1}-1}^\beta - s_{m_{\nu+2}}^\beta\right)\frac{1}{2^{\nu(\gamma-\beta)}}\left(\sum_{k\in I_\nu}s_k^\delta|x_k|\right)^q$$
$$\leq K\sum_{n=m_{\nu+1}-1}^{m_{\nu+2}-1}\left(s_n^\beta - s_{n+1}^\beta\right)s_n^{\gamma-\beta}\left(\sum_{k\in I_\nu}s_k^\delta|x_k|\right)^q .$$

Here we have used $I_\nu \neq \emptyset$ once more: It implies that $m_\nu \leq m_{\nu+1} - 1$, so that for $m_{\nu+1} - 1 \leq n \leq m_{\nu+2} - 1$ we have $1/2^{\nu+2} < s_n \leq 1/2^\nu$, hence $s_n^{\gamma-\beta} \geq K/2^{\nu(\gamma-\beta)}$. Thus we have

$$\left(\frac{1}{2^{\nu\alpha}}\sum_{k\in I_\nu}|x_k|\right)^q \leq K\sum_{n=m_{\nu+1}-1}^{m_{\nu+2}-1}\left(s_n^\beta - s_{n+1}^\beta\right)s_n^{\gamma-\beta}\left(\sum_{k=1}^{n}s_k^\delta|x_k|\right)^q$$

with a constant K independent of ν. It clearly suffices to sum on all $\nu \geq 1$ with $I_\nu \neq \emptyset$. It is then easily seen that the nth term on the right hand side appears at most twice for each n. This shows that (2.4i) implies (2.3) for any $q > 0$.

The proof of the equivalence (2.3)$\Leftrightarrow$(2.4iii) is similar and is therefore omitted. We only remark that in the second part of the proof one has to estimate $1/2^{\nu\beta}$ by $s_{m_\nu}^\beta - s_{m_{\nu-1}-1}^\beta$.

The equivalence of (2.4ii) with (2.3) is obvious once it is noted that (2.4ii) is simply obtained from (2.4i) by substituting β by $-\beta$. The same is true for (2.4iv) and (2.4iii).

To complete the proof of the theorem it suffices to show that in the special case of $\beta = \gamma$ the conditions (2.4i) and (2.4v) are equivalent, as well as (2.4iii) and (2.4vi). This is an immediate consequence of Lemma 3 of **BIII** and the following asymptotic estimates. For $r > -1$ and any non-negative numbers c_k we have

$$\sum_{k=1}^{n} c_k \left(\sum_{j=1}^{k} c_j\right)^r \sim \left(\sum_{j=1}^{n} c_j\right)^{r+1}, \tag{2.6i}$$

and, if in addition $\sum_k c_k < \infty$, then

$$\sum_{k=n}^{\infty} c_k \left(\sum_{j=k}^{\infty} c_j\right)^r \sim \left(\sum_{j=n}^{\infty} c_j\right)^{r+1} \tag{2.6ii}$$

(in the sense that both sides are of the same order as $n \to \infty$).

These estimates, which seem to be folklore, have been used (and proved) by various authors. See, for instance, [2, p. 844] and Lemmas 2 and 3 in **BII**, where one direction in each of these estimates is obtained, the other being trivial. □

The next two theorems treat the case where p or q is infinite.

Theorem 2.3 *Let* $0 < q < \infty$ *and* $\alpha \in \mathbb{R}$. *Then, for any sequence* $x = (x_n)$, *the condition*

$$\sum_{\nu} \left(\frac{1}{2^{\nu\alpha}} \sup_{k \in I_\nu} |x_k|\right)^q < \infty \tag{2.7}$$

is equivalent to each of the following conditions, where $\beta \neq 0, \gamma \neq 0$ *and* δ *are real numbers with* $\gamma/q + \delta = \alpha$:

$$\sum_n (s_n^\beta - s_{n+1}^\beta) s_n^{\gamma-\beta} \left(\sup_{k \le n} s_k^\delta |x_k|\right)^q < \infty \qquad \text{if } \beta > 0, \gamma > 0, \tag{2.8i}$$

$$\sum_n (s_{n+1}^\beta - s_n^\beta) s_n^\gamma s_{n+1}^{-\beta} \left(\sup_{k \le n} s_k^\delta |x_k|\right)^q < \infty \qquad \text{if } \beta < 0, \gamma > 0, \tag{2.8ii}$$

$$\sum_n (s_n^\beta - s_{n-1}^\beta) s_n^{\gamma-\beta} \left(\sup_{k \ge n} s_k^\delta |x_k|\right)^q < \infty \qquad \text{if } \beta < 0, \gamma < 0, \tag{2.8iii}$$

$$\sum_n (s_{n-1}^\beta - s_n^\beta) s_n^\gamma s_{n-1}^{-\beta} \left(\sup_{k \ge n} s_k^\delta |x_k|\right)^q < \infty \qquad \text{if } \beta > 0, \gamma < 0. \tag{2.8iv}$$

Remark 2.4 For an abstract sequence (m_ν) (or (s_n)) and $\alpha \neq 0$ the simplest conditions equivalent to (2.7) are $(\beta = \gamma = \alpha q, \delta = 0)$:

$$\text{(2.9i)} \qquad \sum_n (s_n^{\alpha q} - s_{n+1}^{\alpha q}) \left(\sup_{k \le n} |x_k| \right)^q < \infty \qquad \text{if} \quad \alpha > 0,$$

$$\text{(2.9ii)} \qquad \sum_n (s_n^{\alpha q} - s_{n-1}^{\alpha q}) \left(\sup_{k \ge n} |x_k| \right)^q < \infty \qquad \text{if} \quad \alpha < 0.$$

We refer to Remark 2.2(i) for the interpretation of s_0. We do not know if, in analogy to Theorem 2.1, there are other equivalent conditions that avoid differences.

Proof of Theorem 2.3. As in the proof of Theorem 2.1 we need only consider the case $q = 1$, and we need only prove the equivalence of (2.7) with (2.8i) and (2.8iii).

(2.7)⇒(2.8iii). Fix $\nu \ge 1$. Using (2.1) we obtain

$$\begin{aligned}
\sum_{n \in I_\nu} (s_n^\beta - s_{n-1}^\beta) s_n^{\gamma-\beta} \sup_{k \ge n} s_k^\delta |x_k| &\le K \sum_{n \in I_\nu} (s_n^\beta - s_{n-1}^\beta) \frac{1}{2^{\nu(\gamma-\beta)}} \sup_{k \ge m_\nu} s_k^\delta |x_k| \\
&\le K (s_{m_{\nu+1}-1}^\beta - s_{m_\nu - 1}^\beta) \frac{1}{2^{\nu(\gamma-\beta)}} \sup_{\mu \ge \nu} \sup_{k \in I_\mu} \frac{1}{2^{\mu\delta}} |x_k| \\
&\le K \frac{1}{2^{\nu\gamma}} \sup_{\mu \ge \nu} \frac{1}{2^{\mu\delta}} \sup_{k \in I_\mu} |x_k| \\
&= K \sup_{\mu \ge \nu} \frac{1}{2^{(\nu-\mu)\gamma}} X_\mu,
\end{aligned}$$

where we set $X_\mu = 1/2^{\mu\alpha} \sup_{k \in I_\mu} |x_k|$. Summing on ν we have

$$\text{(2.10)} \qquad \sum_{n=m_1}^{\infty} (s_n^\beta - s_{n-1}^\beta) s_n^{\gamma-\beta} \sup_{k \ge n} s_k^\delta |x_k| \le K \sum_\nu \sup_{\mu \ge \nu} \frac{1}{2^{(\nu-\mu)\gamma}} X_\mu.$$

For any $N \in \mathbb{N}$ we observe that, since $\gamma < 0$,

$$\begin{aligned}
\sum_{\nu=0}^{N} \sup_{\mu \ge \nu} \frac{1}{2^{(\nu-\mu)\gamma}} X_\mu &\le \sum_{\nu=0}^{N-1} \left(2^\gamma \sup_{\mu \ge \nu+1} \frac{1}{2^{(\nu+1-\mu)\gamma}} X_\mu + X_\nu \right) + \sup_{\mu \ge N} \frac{1}{2^{(N-\mu)\gamma}} X_\mu \\
&\le 2^\gamma \sum_{\nu=0}^{N} \sup_{\mu \ge \nu} \frac{1}{2^{(\nu-\mu)\gamma}} X_\mu + \sum_{\nu=0}^{N-1} X_\nu + \sup_{\mu \ge N} X_\mu.
\end{aligned}$$

From this we deduce after letting $N \to \infty$ that

$$\text{(2.11)} \qquad \sum_\nu \sup_{\mu \ge \nu} \frac{1}{2^{(\nu-\mu)\gamma}} X_\mu \le \frac{1}{1 - 2^\gamma} \sum_\nu X_\nu,$$

so that (2.8iii) follows from (2.10) and (2.11) under the assumption of (2.7).

(2.8iii)⇒(2.7). With the obvious changes the proof is the same as that of (2.4iii) ⇒(2.3). The similar (and slightly simpler) proof of (2.7)⇔(2.8i) is omitted. □

Theorem 2.5 *Let $0 < p < \infty$ and $\alpha \in \mathbb{R}$. Then, for any sequence $\boldsymbol{x} = (x_n)$, the condition*

$$\sup_\nu \frac{1}{2^{\nu\alpha}} \left(\sum_{k \in I_\nu} |x_k|^p \right)^{1/p} < \infty \tag{2.12}$$

is equivalent to each of the following conditions, where $\gamma \neq 0$ and δ are real numbers with $\gamma + \delta/p = \alpha$:

$$\sup_n s_n^\gamma \left(\sum_{k=1}^{n} s_k^\delta |x_k|^p \right)^{1/p} < \infty \qquad \text{if } \gamma > 0, \tag{2.13i}$$

$$\sup_n s_n^\gamma \left(\sum_{k=n}^{\infty} s_k^\delta |x_k|^p \right)^{1/p} < \infty \qquad \text{if } \gamma < 0. \tag{2.13ii}$$

In addition, for real numbers $\gamma \neq 0, \delta, \eta$ and σ with $\gamma + \delta/p = \alpha$ and $\rho, \sigma > 0$, (2.12) is equivalent to

$$\sum_{k=1}^{n} \left(s_k^{\rho\gamma} - s_{k+1}^{\rho\gamma} \right) \left(\sum_{j=1}^{k} s_j^\delta |x_j|^p \right)^{(\rho+\sigma)/p} = O\left(\sum_{k=1}^{n} s_k^\delta |x_k|^p \right)^{\sigma/p} \quad \text{if } \gamma > 0, \tag{2.13iii}$$

$$\sum_{k=n}^{\infty} \left(s_k^{\rho\gamma} - s_{k-1}^{\rho\gamma} \right) \left(\sum_{j=k}^{\infty} s_j^\delta |x_j|^p \right)^{(\rho+\sigma)/p} = O\left(\sum_{k=n}^{\infty} s_k^\delta |x_k|^p \right)^{\sigma/p} \quad \text{if } \gamma < 0. \tag{2.13iv}$$

Remark 2.6 The simplest conditions equivalent to (2.12) in case $\alpha \neq 0$ are

$$\sup_n s_n^\alpha \left(\sum_{k=1}^{n} |x_k|^p \right)^{1/p} < \infty \qquad \text{if } \alpha > 0, \tag{2.14i}$$

$$\sup_n s_n^\alpha \left(\sum_{k=n}^{\infty} |x_k|^p \right)^{1/p} < \infty \qquad \text{if } \alpha < 0. \tag{2.14ii}$$

Proof of Theorem 2.5. We note that by an obvious change of variables and parameters we can assume that $p = 1$.

(2.12)⇒(2.13i). Let $\nu \geq 1$. Then we have for $n \in I_\nu$, using (2.1),

$$\begin{aligned} s_n^\gamma \sum_{k=1}^{n} s_k^\delta |x_k| &\leq K \frac{1}{2^{\nu\gamma}} \sum_{\mu=0}^{\nu} \sum_{k \in I_\mu} \frac{1}{2^{\mu\delta}} |x_k| \\ &\leq K \sum_{\mu=0}^{\nu} \frac{1}{2^{\mu\alpha}} \frac{1}{2^{\gamma(\nu-\mu)}} \sum_{k \in I_\mu} |x_k| \\ &\leq K \sup_\mu \left(\frac{1}{2^{\mu\alpha}} \sum_{k \in I_\mu} |x_k| \right) \sum_{\mu=0}^{\nu} \frac{1}{2^{\gamma(\nu-\mu)}}. \end{aligned}$$

This implies (2.13i) since $\gamma > 0$.

(2.13i)⇒(2.12). For $\nu \geq 1$ with $I_\nu \neq \emptyset$ we obtain, again using (2.1):

$$\begin{aligned} \frac{1}{2^{\nu\alpha}} \sum_{k \in I_\nu} |x_k| &= \frac{1}{2^{\nu\gamma}} \sum_{k \in I_\nu} \frac{1}{2^{\nu\delta}} |x_k| \\ &\leq K s_{m_{\nu+1}-1}^\gamma \sum_{k \in I_\nu} s_k^\delta |x_k| \\ &\leq K s_n^\gamma \sum_{k=1}^{n} s_k^\delta |x_k| \end{aligned}$$

with $n = m_{\nu+1} - 1$. Condition (2.12) now follows.

(2.12)⇔(2.13ii) is proved similarly.

(2.13i)⇔(2.13iii). We prove this equivalence by applying the just now established equivalence of (2.13i) and (2.14ii) (!) to certain new sequences (X_n) and (S_n) and new parameters Γ, Δ and A; we use upper-case letters to avoid confusion. Let $\gamma, \rho, \sigma > 0$ and $\delta \in \mathbb{R}$ with $\gamma + \delta = \alpha$. We define, for $n \geq 1$,

$$T_n = \sum_{k=1}^{n} s_k^\delta |x_k|$$

and

$$X_n = s_n^{\rho\gamma} - s_{n+1}^{\rho\gamma}.$$

If (T_n) is bounded, then (2.13i) and (2.13iii) both hold, so that nothing is to be proved. Hence we assume that $T_n \to \infty$. Also, after possibly shifting indices, we may assume that $T_n \neq 0$ for all n. We then set

$$S_n = T_n^{-1},$$

which gives us a monotonic positive sequence with $S_n \to 0$. With these sequences $(S_n), (X_n)$ and the parameters $\Gamma = \sigma, \Delta = -(\rho + \sigma)$ and $A = -\rho$ (with $p = 1$) the equivalent conditions (2.13i) and (2.14ii) read

$$\sup_n \left(\sum_{k=1}^{n} s_k^\delta |x_k| \right)^{-\sigma} \sum_{k=1}^{n} \left(\sum_{j=1}^{k} s_j^\delta |x_j| \right)^{\rho+\sigma} (s_k^{\rho\gamma} - s_{k+1}^{\rho\gamma}) < \infty$$

and

$$\sup_n \left(\sum_{k=1}^{n} s_k^\delta |x_k| \right)^{\rho} \sum_{k=n}^{\infty} (s_k^{\rho\gamma} - s_{k+1}^{\rho\gamma}) < \infty,$$

which give (2.13iii) and (2.13i), respectively, for the sequences (s_n) and (x_n).

(2.13ii)$\Leftrightarrow$(2.13iv) is proved similarly. □

For the sake of completeness we include the almost trivial case of $p = q = \infty$. The proof follows the same lines as that of Theorem 2.5.

Theorem 2.7 *Let $\alpha \in \mathbb{R}$. Then, for any sequence $\boldsymbol{x} = (x_n)$, the condition*

$$\sup_\nu \frac{1}{2^{\nu\alpha}} \sup_{k \in I_\nu} |x_k| < \infty \tag{2.15}$$

is equivalent to the following conditions, where γ and δ are real numbers with $\gamma + \delta = \alpha$:

(2.16i) $\quad \sup_n s_n^\gamma \sup_{k \le n} s_k^\delta |x_k| < \infty \qquad$ *if* $\gamma \geq 0$,

(2.16ii) $\quad \sup_n s_n^\gamma \sup_{k \ge n} s_k^\delta |x_k| < \infty \qquad$ *if* $\gamma \leq 0$.

Remark 2.8 We note that here the case $\gamma = 0$ need not be excluded as in Theorem 2.5. Indeed, this γ gives us the simplest condition equivalent to (2.15):

$$\sup_n s_n^\alpha |x_n| < \infty. \tag{2.17}$$

3 Transformation from Section Form into Block Form

These transformations are now easily found by reading the results of the previous section "backwards". Throughout this section, let $\boldsymbol{a} = (a_n)$ be a fixed sequence of non-negative numbers. In order to avoid trivialities we assume that $\boldsymbol{a}$ is not a finite sequence.

Theorem 3.1 *Let $\boldsymbol{x} = (x_n)$ be any sequence.*

(a) *Let $0 < p, q < \infty$. Assume that $\sum_k a_k^q < \infty$, and define an index sequence $\boldsymbol{m}$ by letting m_ν be the least n such that $(\sum_{k=n}^{\infty} a_k^q)^{1/q} \le 1/2^\nu$ $(\nu \ge 1)$. Then*

$$\sum_n \left[a_n \left(\sum_{k=1}^n |x_k|^p \right)^{1/p} \right]^q < \infty$$

is equivalent to

$$\sum_\nu \left[\frac{1}{2^\nu} \left(\sum_{k \in I_\nu} |x_k|^p \right)^{1/p} \right]^q < \infty.$$

(b) *Let $0 < q < \infty$. Assume that $\sum_k a_k^q < \infty$, and define $\boldsymbol{m}$ as in* (a). *Then*

$$\sum_n \left(a_n \sup_{k \le n} |x_k| \right)^q < \infty$$

is equivalent to

$$\sum_\nu \left(\frac{1}{2^\nu} \sup_{k \in I_\nu} |x_k| \right)^q < \infty.$$

(c) *Let $0 < p < \infty$. Assume that $a_k \to 0$, and define an index sequence $\boldsymbol{m}$ by letting m_ν be the least n such that $\sup_{k \ge n} a_k \le 1/2^\nu$ $(\nu \ge 1)$. Then*

$$\sup_n a_n \left(\sum_{k=1}^n |x_k|^p \right)^{1/p} < \infty$$

is equivalent to

$$\sup_\nu \frac{1}{2^\nu} \left(\sum_{k \in I_\nu} |x_k|^p \right)^{1/p} < \infty.$$

(d) *Assume that $a_k \to 0$, and define $\boldsymbol{m}$ as in* (c). *Then*

$$\sup_n a_n \sup_{k \le n} |x_k| < \infty$$

is equivalent to

$$\sup_\nu \frac{1}{2^\nu} \sup_{k \in I_\nu} |x_k| < \infty.$$

Proof. (a) Setting $s_n = \left(\sum_{k=n}^{\infty} a_k^q\right)^{1/q}$ we see that the sequences $\boldsymbol{s}$ and $\boldsymbol{m}$ satisfy the assumptions of Section 2 and are correlated in the sense introduced there (compare (2.2)). Now, with $\alpha = 1, \beta = \gamma = q$ and $\delta = 0$, the equivalence of (2.3) and (2.4i) of Theorem 2.1 gives (a).

(b) follows in the same way from Theorem 2.3.

Similary, (c) and (d) can be deduced from Theorems 2.5 and 2.7 when we set $s_n = \sup_{k\geq n} a_k$. One has to note in addition that for any non-negative, non-decreasing sequence (c_n) we have $\sup_n(\sup_{k\geq n} a_k)c_n = \sup_n a_n c_n$. □

We turn to the companion result for tail sections.

Theorem 3.2 *Let $\boldsymbol{x} = (x_n)$ be any sequence.*

(a) *Let $0 < p, q < \infty$. Assume that $\sum_k a_k^q = \infty$, and define an index sequence $\boldsymbol{m}$ by letting m_ν be the least n such that $(\sum_{k=1}^{n} a_k^q)^{1/q} \geq 2^\nu$ $(\nu \geq 1)$. Then*

$$\sum_n \left[a_n \left(\sum_{k=n}^{\infty} |x_k|^p\right)^{1/p}\right]^q < \infty$$

is equivalent to

$$\sum_\nu \left[2^\nu \left(\sum_{k\in I_\nu} |x_k|^p\right)^{1/p}\right]^q < \infty.$$

(b) *Let $0 < q < \infty$. Assume that $\sum_k a_k^q = \infty$, and define $\boldsymbol{m}$ as in* (a). *Then*

$$\sum_n \left(a_n \sup_{k\geq n} |x_k|\right)^q < \infty$$

is equivalent to

$$\sum_\nu \left(2^\nu \sup_{k\in I_\nu} |x_k|\right)^q < \infty.$$

(c) *Let $0 < p < \infty$. Assume that $\sup_k a_k = \infty$, and define an index sequence $\boldsymbol{m}$ by letting m_ν be the least n such that $\sup_{k\leq n} a_k \geq 2^\nu$. Then*

$$\sup_n a_n \left(\sum_{k=n}^{\infty} |x_k|^p\right)^{1/p} < \infty$$

is equivalent to

$$\sup_\nu 2^\nu \left(\sum_{k\in I_\nu} |x_k|^p\right)^{1/p} < \infty.$$

(d) *Assume that* $\sup_k a_k = \infty$, *and define* m *as in* (c). *Then*

$$\sup_n a_n \sup_{k \geq n} |x_k| < \infty$$

is equivalent to

$$\sup_\nu 2^\nu \sup_{k \in I_\nu} |x_k| < \infty.$$

Proof. After shifting indices, if necessary, we see that we may assume $a_1 > 0$. (a) and (b) then follow from Theorems 2.1 und 2.3 upon setting $s_n = \left(\sum_{k=1}^n a_k^q\right)^{-1/q}$ and $\alpha = -1, \beta = \gamma = -q$, $\delta = 0$. In the same way (c) and (d) follow from Theorems 2.5 and 2.7 with $s_n = (\sup_{k \leq n} a_k)^{-1}$ if one notes in addition that $\sup_n (\sup_{k \leq n} a_k) c_n = \sup_n a_n c_n$ for any non-negative, non-increasing sequence (c_n). □

4 Comments

In this section we want to discuss various points pertaining to the results of this chapter.

Constants

The main results in the previous two sections state that for certain (extended quasi-semi)norms $\|\cdot\|_{\text{block}}$ and $\|\cdot\|_{\text{section}}$ in block and section form we have

$$\forall\, x \quad \|x\|_{\text{block}} < \infty \quad \Leftrightarrow \quad \|x\|_{\text{section}} < \infty. \tag{4.1}$$

Now, if $s_0 = \infty$ in Theorems 2.1 and 2.3 (cf. Remark 2.2(i)) and if $a_1 > 0$ in Theorem 3.2, then these norms are positive definite. In that case, by a well-known reasoning based on the closed graph theorem, (4.1) is equivalent to the more quantitative statement that there are constants K and K' such that

$$\forall\, x \quad K\|x\|_{\text{block}} \leq \|x\|_{\text{section}} \leq K'\|x\|_{\text{block}}. \tag{4.2}$$

In Section 2 these constants depend on the sequence s (or m) and on the various parameters $p, \alpha, \ldots$. What we want to emphasise here is that the dependence on s is rather loose: If we consider only sequences s with $s_1 (= \sup_n s_n) \leq M$ for a fixed $M > 0$, then K and K' can be taken to be independent of s. A corresponding remark then applies to our results in Section 3 (see the definition of s in the proofs there). We shall later demonstrate the usefulness of (4.2) in this stronger form, see Proposition 13.4.

In order to justify our claim one need only extend the proofs in Section 2 to cover the case $\nu = 0$ as well and trace the constants K carefully. We leave the rather tedious details to the reader.

Extensions

The base 2 appearing in norms of block form is of no importance. All the results in the previous two sections remain true if 2 is replaced by any other base $\rho > 1$. This extra flexibility will be needed in one of our applications (see Example 13.2). Even more generally one might think of replacing the coefficients $1/2^{\nu\alpha}$ by $\varphi(2^\nu)$ for some function φ. Norms in block form of this kind have already appeared in the literature, see for example [16] or [65]. Some extension of our results in this direction is surely possible. We shall not pursue this here.

A different extension is in reality an application: The results remain true for vector-valued sequences (x_k) with x_k belonging to certain Banach spaces X_k if $|x_k|$ is replaced by $\|x_k\|_{X_k}$. This is of relevance, for instance, in connection with Problem 4 of Lee [56].

Previous work

Scattered in the literature one finds a large number of equivalence results involving norms with dyadic blocks. They are often hidden as lemmas or inside some proof and needed for a particular purpose. We only refer to [32, Lemma 1].

The case of non-dyadic blocks has attracted far less attention. Some particular equivalence results in this direction have been obtained by Okuyama and Tsuchikura [80] and in several papers of Leindler, see [59, 65] and the literature cited therein.

A very interesting investigation parallel to the one in this chapter but with a different emphasis is due to Totik and Vincze [97] and Leindler [65]. They assume that a norm in section form and another one in block form are given and characterise when they are equivalent. Such results allow one to check equivalence once an appropriate translation has been found. Thus the equivalence of (2.3) with (2.4iii/iv) in Theorem 2.1 can also be proved using Theorem 1 of [97]. The equivalence of (2.3) with (2.4i/ii) corresponds to Theorem 1 of [65]. However, the condition on B_m in the latter result is not in general necessary (its necessity is only proved in a special case). A counterexample can be constructed using Theorem 2.1 above.

An interesting aspect of the work of Totik-Vincze and Leindler is that in their norms of block form they allow any monotonic sequence of coefficients instead of $1/2^{\nu\alpha}$. As far as this special case is considered, however, we shall show in the Appendix that the results in this chapter in turn lead to a proof and to an extension of the results of Totik-Vincze and Leindler.

Applicability

In the remainder of these notes we shall apply the results of this chapter to a variety of problems. Common to most of these applications is the following procedure: Given a problem that is formulated in terms of norms in section form we use Section 3 to translate them into block form. We then solve this new, and usually much simpler, problem. Very often this solution involves a norm in block form, which we then re-translate into section form with the help of Section 2.

Under particular circumstances, however, one can read and apply our results also in a different direction. Theorem 2.1, for instance, can be viewed as ascertaining the equivalence of the conditions (2.4i) - (2.4vi) with possibly different β, γ and δ. Condition (2.3) then only acts as a catalyst.

We give one example here; for another one see Theorem 13.1 below. Let $0 < p, q < \infty$ and define s by $\frac{1}{s} = \frac{1}{p} + \frac{1}{q}$. Furthermore, let (a_n) be any sequence of non-negative terms with $a_1 > 0$, and set $A_n = \sum_{k=1}^n a_k$. As a consequence of some other results Bennett has concluded that for any sequence x the conditions

$$\text{(4.3i)} \qquad \sum_n a_n \left(\sum_{k=n}^{\infty} \frac{|x_k|^s}{A_k} \right)^{p/s} < \infty$$

and

$$\text{(4.3ii)} \qquad \sum_n |x_n|^s \left(\frac{1}{A_n} \sum_{k=1}^{n} |x_k|^s \right)^{p/q} < \infty$$

are equivalent (**BIV**, p. 24). He asked for a simple, direct proof of this fact. Now, if in Theorem 2.1 we assign the parameters p, q, s_n and α the values $s, p, 1/A_n$ and $1/q$, then condition (2.4iii) becomes (4.3i) if we take $\beta = \gamma = -1$ and $\delta = 1$, and (2.4v) becomes (4.3ii) under $\gamma = p/q$ and $\delta = 0$. Thus Theorem 2.1 implies the desired equivalence if we assume that $A_n \to \infty$ (hence $s_n \to 0$), the only case of interest in **BIV**. We leave it to the reader to decide if the proof of the equivalence in Theorem 2.1 can be called simple.

To end this chapter we spell out the most prominent particular case of our results, the renorming of the Cesàro sequence space ces (p), thus confirming a claim made in the Introduction.

Theorem 4.1 *For $1 < p < \infty$ the space* $\mathrm{ces}(p)$ *consists of all sequences $\boldsymbol{x}$ for which*

$$\|\boldsymbol{x}\| := \left(\sum_{\nu=0}^{\infty} 2^{\nu(1-p)} \left(\sum_{k=2^\nu}^{2^{\nu+1}-1} |x_k|\right)^p\right)^{1/p} < \infty,$$

and this yields an equivalent norm on $\mathrm{ces}(p)$.

Proof. Instead of Theorem 3.1, which gives different and more difficult blocks, we apply Theorem 2.1. Taking $s_n = 1/n$ with $\alpha = (p-1)/p, \beta = -1, \gamma = p-1$ and $\delta = 0$ we deduce that $\|\boldsymbol{x}\| < \infty$ if and only if

$$\sum_n \frac{1}{n+1} \frac{1}{n^{p-1}} \left(\sum_{k=1}^{n} |x_k|\right)^p < \infty,$$

which is equivalent to $\boldsymbol{x} \in \mathrm{ces}(p)$. An application of the closed graph theorem then implies the equivalence of the norms. □

There has been an earlier attempt to write $\mathrm{ces}(p)$ in block form. In [67], Lim considered the space

$$\left\{\boldsymbol{x} : \sum_{\nu=0}^{\infty} 2^{-\nu p} \left(\sum_{k=2^\nu}^{2^{\nu+1}-1} |x_k|\right)^p < \infty\right\} \quad (1 < p < \infty).$$

By the above result this space differs from $\mathrm{ces}(p)$, see also [8]. In fact, by Theorem 2.1, Lim's space coincides with

$$\left\{\boldsymbol{x} : \sum_n \frac{1}{n} \left(\frac{1}{n} \sum_{k=1}^{n} |x_k|\right)^p < \infty\right\}.$$

Chapter II

The Sequence Spaces $c(\boldsymbol{a}, p, q)$ and $d(\boldsymbol{a}, p, q)$

5 The Spaces c and d, and $l(p, q)$

In Chapter I we were primarily concerned with (extended quasi-semi)norms in section and in block form. These lead naturally to sequence spaces that we shall now introduce.

Let $\boldsymbol{a} = (a_n)$ be any sequence of non-negative terms. Then, for $0 < p, q \leq \infty$, we define the following spaces:

$$c(\boldsymbol{a}, p, q) = \left\{ \boldsymbol{x} : \sum_n \left[a_n \left(\sum_{k=1}^{n} |x_k|^p \right)^{1/p} \right]^q < \infty \right\}$$

and

$$d(\boldsymbol{a}, p, q) = \left\{ \boldsymbol{x} : \sum_n \left[a_n \left(\sum_{k=n}^{\infty} |x_k|^p \right)^{1/p} \right]^q < \infty \right\},$$

with the usual modifications if p or q is infinite. Not all sequences $\boldsymbol{a}$, however, lead to new spaces. Before we proceed we single out, once and for all, certain $\boldsymbol{a}$ for which the spaces c and d reduce to, in some sense, trivial spaces.

Proposition 5.1 *Let $0 < p, q \leq \infty$. Then*

$$\text{(a)} \quad \begin{aligned} c(\boldsymbol{a},p,q) &= \{0\} && \text{if} \quad \boldsymbol{a} \notin l^q, \\ &= l^p && \text{if} \quad \boldsymbol{a} \in l^\infty \setminus c_0 \text{ and } q = \infty, \\ &= \omega && \text{if} \quad \boldsymbol{a} \in \varphi. \end{aligned}$$

$$\text{(b)} \quad d(\boldsymbol{a},p,q) = l^p \quad \text{if} \quad \boldsymbol{a} \in l^q.$$

One might just as well put $d(\boldsymbol{0},p,q) = \omega$. We have here chosen to define it as l^p in accordance with our convention at the end of the Introduction.

We shall refer to the sequences $\boldsymbol{a}$ described above as the *trivial* sequences. Of course, the triviality of $\boldsymbol{a}$ depends on whether we consider a space c or d and on the parameters p and q. We are clearly primarily interested in the non-trivial $\boldsymbol{a}$, and some of our results will only hold for these.

Now, to continue, assume that $\boldsymbol{a}$ is non-trivial. Then we endow the sequence spaces c and d with their natural quasi-norms (cf. [53, §15.10], [49])

$$\|\boldsymbol{x}\|_{c(\boldsymbol{a},p,q)} = \left(\sum_n \left[a_n \left(\sum_{k=1}^n |x_k|^p \right)^{1/p} \right]^q \right)^{1/q}$$

and

$$\|\boldsymbol{x}\|_{d(\boldsymbol{a},p,q)} = \left(\sum_n \left[a_n \left(\sum_{k=n}^\infty |x_k|^p \right)^{1/p} \right]^q \right)^{1/q} + \|\boldsymbol{x}\|_{l^p},$$

with modifications if p or q is infinite. For d, the term $\|\boldsymbol{x}\|_{l^p}$ may (and will) be dropped if $a_1 > 0$.

The notation employed here was suggested by two special cases, the Cesàro sequence space $\mathrm{ces}(p) = c(1/n, 1, p)$ and Bennett's spaces $d(\boldsymbol{a},p) = d(a_n^{1/p}, \infty, p)$, see §3 in **BIV**. Many other spaces of the literature are covered by the spaces c and d. We only mention Jagers' spaces $b_p = c(\beta_n, 1, p)$ for some (β_n), see [44], and Bennett's second fundamental class $g(\boldsymbol{a},p) = c(A_n^{-1/p}, p, \infty)$, see **BIV**, §3. Further examples will appear in the course of these notes.

As counterpart to the spaces c and d we have the sequence spaces that are defined by norms in block form. Let $\boldsymbol{m} = (m_\nu)$ be any index sequence. Then, for $0 < p, q \leq \infty$, we define

$$l(\boldsymbol{m},p,q) = \left\{ \boldsymbol{x} : \sum_\nu \left(\sum_{k=m_\nu}^{m_{\nu+1}-1} |x_k|^p \right)^{q/p} < \infty \right\},$$

with the usual modifications if p or q is infinite. These spaces are endowed with their obvious quasi-norms. Again, for some index sequences the spaces reduce to well-known spaces. Since all quasi-norms $\|\cdot\|_{l^p_n}, 0 < p \leq \infty$, are equivalent for fixed n, we have the following.

Proposition 5.2 *Let $0 < p, q \leq \infty$. Then*

$$l(\boldsymbol{m}, p, q) = l^q \quad \textit{if } (m_{\nu+1} - m_\nu)_\nu \textit{ is bounded.}$$

So far, the spaces $l(\boldsymbol{m}, p, q)$ have only appeared in the literature in their dyadic variety, denoted as

$$l(p, q) = l(2^\nu, p, q).$$

These spaces were introduced by Hedlund [41] and subsequently studied by Kellogg [51] and others, see [50], [24], [77], [78], [70], [48]. The extension of their results from dyadic to general blocks causes hardly any problem.

We remark that the more general norms in block form appearing in Chapter I, $\left(\sum_\nu [1/2^{\nu\alpha}(\sum_{k\in I_\nu} |x_k|^p)^{1/p}]^q\right)^{1/q}$, lead to a larger class of sequence spaces. In the special case of dyadic blocks they were introduced as Besov sequence spaces $b^\alpha_{p,q}$ by Pietsch [82]. However, after taking the factor $1/2^{\nu\alpha}$ inside the interior sum one sees that these spaces are weighted $l(\boldsymbol{m}, p, q)$-spaces. Thus we do not need this additional generality here.

To end this section we summarise the results of Section 3 as follows.

Observation 5.3 *Every sequence space $c(\boldsymbol{a}, p, q)$ and every $d(\boldsymbol{a}, p, q)$ is a weighted $l(\boldsymbol{m}, p, q)$, if $\boldsymbol{a}$ is non-trivial.*

It is interesting to note that the restrictions on $\boldsymbol{a}$ that were forced upon us in Section 3 coincide with the distinction between trivial and non-trivial $\boldsymbol{a}$ introduced above. By Proposition 5.1 the above observation remains true for all $\boldsymbol{a} \in l^q \setminus \varphi$ in case of space c and, indeed, for all $\boldsymbol{a}$ in case of space d.

6 Structure of the Spaces c and d

Some first structural properties of the sequence spaces $c(\boldsymbol{a}, p, q)$ and $d(\boldsymbol{a}, p, q)$ follow rather directly from their definitions. We only consider non-trivial $\boldsymbol{a}$.

- They are quasi-Banach spaces (cf. [49]), and Banach spaces if $p, q \geq 1$.

- They are K-spaces, which means that all coordinate functionals $\boldsymbol{x} \mapsto x_n$ ($n \in \mathbb{N}$) are continuous. Thus we have BK-spaces for $p, q \geq 1$ and Fréchet K-spaces for general p, q.

- For $q < \infty$ they have the AK-property, which means that for every sequence $\boldsymbol{x}$ in the space its sections $P_n\boldsymbol{x} = (x_1, x_2, \dots, x_n, 0, 0, \dots)$ converge to $\boldsymbol{x}$ in the topology of the space. This is the same as saying that the unit sequences $\boldsymbol{e_n} = (\delta_{nk})_k$ form a Schauder basis.

If a K-space E does not have the AK-property, its subspace E_{AK} of all sequences $\boldsymbol{x}$ with $P_n\boldsymbol{x} \to \boldsymbol{x}$ in E often comes into play. The following is easily established.

Proposition 6.1 *Let $0 < p \leq \infty$ and assume that $\boldsymbol{a}$ is non-trivial (that is, $\boldsymbol{a} \in c_0 \setminus \varphi$ for c, $\boldsymbol{a} \notin l^\infty$ for d). Then*

$$c(\boldsymbol{a}, p, \infty)_{AK} = \left\{ \boldsymbol{x} : a_n \left(\sum_{k=1}^{n} |x_k|^p \right)^{1/p} \to 0 \quad as \quad n \to \infty \right\},$$

$$d(\boldsymbol{a}, p, \infty)_{AK} = \left\{ \boldsymbol{x} : a_n \left(\sum_{k=n}^{\infty} |x_k|^p \right)^{1/p} \to 0 \quad as \quad n \to \infty \right\},$$

with the usual interpretation if $p = \infty$.

In order to derive deeper structural properties we shall now apply the results of Section 3 (via Observation 5.3). Every space $l(\boldsymbol{m}, p, q)$ is reflexive for $1 \leq p \leq \infty$ if $1 < q < \infty$ but not for $q = 1$ or $q = \infty$ (as l^q-sum of reflexive spaces, see also [77, Corollaries 1 and 2]), and it is known to be perfect for $1 \leq p, q \leq \infty$ ([51]). We recall that a sequence space is called *perfect* if it coincides with its Köthe-bidual (see Section 7). As a consequence we obtain the following.

Theorem 6.2 *For each non-trivial sequence $\boldsymbol{a}$ the sequence spaces $c(\boldsymbol{a}, p, q)$ and $d(\boldsymbol{a}, p, q)$ are reflexive for $1 \leq p \leq \infty$ and $1 < q < \infty$ but not for $q = 1$ or ∞; they are perfect for $1 \leq p, q \leq \infty$.*

The first part of the theorem extends a result of Jagers [44] who considered the spaces $c(\boldsymbol{a}, p, q)$ for $p = 1$. In particular, we can confirm that the Cesàro sequence space ces(p) is reflexive and perfect for $1 < p < \infty$.

Remark 6.3 At this point we digress briefly. We want to add that the structural properties of the spaces c and d that we have studied so far can also be ascertained for a much larger class of spaces. To this end let $A = (a_{nk})$ be any (infinite) matrix with non-negative entries, and let $0 < p, q \leq \infty$. Then we define the sequence space

$$A(p,q) = \left\{ \boldsymbol{x} : \sum_n \left[\sum_k (a_{nk}|x_k|)^p \right]^{q/p} < \infty \right\},$$

with the usual modifications if p or q is infinite. For $p = 1$ this class of spaces was studied by Johnson and Mohapatra (in several papers, see for example [45]) and by Bennett (**BIV**, §17), among others. For $q = \infty$ they have appeared as spaces of strongly bounded sequences with index p in the papers of Maddox and his students (see [73], [71]). Of course, spaces with particular A abound in the literature. In the special case of lower- or upper-triangular factorable matrices which we shall study in Sections 9 and 10 we obtain $c(\boldsymbol{a}, p, q)$ and $d(\boldsymbol{a}, p, q)$, possibly with weights.

Now, in order to avoid trivalities let us assume that all columns of A belong to l^p and that none of them vanishes. We then endow $A(p, q)$ with its natural quasi-norm. As a result, these spaces become quasi-Banach spaces and K-spaces, in particular BK-spaces for $p, q \geq 1$, which have the AK-propertiy for $q < \infty$. It is less immediate that they are reflexive if A is row-finite and $1 \leq p \leq \infty$, $1 < q < \infty$, and that they are perfect for $1 \leq p, q \leq \infty$. This can be proved as in **BIV**, 17.18.

Bennett (**BIV**, 15.13) has shown that the Cesàro sequence space ces(p) is distinct, as a Banach space, from each l^q and from each Orlicz space. This has led him to conclude that "their Banach space structure cannot be readily discerned from that of some previously studied class" (**BIV**, p. 7). We have already seen that this is not quite the case: ces(p) is a weighted $l(1, p)$.

More generally, every space c or d (for non-trivial $\boldsymbol{a}$) is identical to some weighted $l(\boldsymbol{m}, p, q)$-space. We shall now show that, with regard to the isomorphic structure of these spaces, we need only consider dyadic blocks. This will follow from the next result.

Proposition 6.4 *Let* $0 < p, q \leq \infty$, *and let* $\boldsymbol{m} = (m_\nu)$ *be an index sequence. Then*

$$\begin{aligned} l(\boldsymbol{m}, p, q) &\cong l(p, q) \quad \text{if} \quad (m_{\nu+1} - m_\nu)_\nu \text{ is unbounded,} \\ &= l^q \quad \text{if} \quad (m_{\nu+1} - m_\nu)_\nu \text{ is bounded.} \end{aligned}$$

Proof. In view of Proposition 5.2 we may assume that $(m_{\nu+1} - m_\nu)_\nu$ is unbounded. A moment's reflection will reveal that one can find new index sequences (m'_ν) and (n_ν) with the following properties: Every point m_ν is some m'_μ, every 2^ν is some n_μ; between any two points m_ν and $m_{\nu+1}$ there is at most one new m'_μ, between any two points 2^ν and $2^{\nu+1}$ there is at most one new n_μ; and there is a bijection between the sets of intervals $[m'_\nu, m'_{\nu+1})$ and $[n_\mu, n_{\mu+1})$ so that corresponding intervals have equal length. This then implies that $l(\boldsymbol{m}, p, q) \cong l(\boldsymbol{m'}, p, q) \cong l(\boldsymbol{n}, p, q) \cong l(p, q)$. □

Recently we learnt that, in the Banach space case of $p, q \geq 1$, W. Lusky [72, Lemma 4.1] has obtained the same result using the concept of Banach-Mazur distance.

Since $l^q = l(q, q)$, the proposition tells us that every $l(\boldsymbol{m}, p, q)$ is isomorphic to some (dyadic) $l(\tilde{p}, q)$. We remark that even some of the spaces $l(p, q)$ coincide isomorphically: it was shown by Pełczyński, using Khintchine's inequality, that $l(2, q) \cong l^q$ for $1 < q < \infty$ (see [68, Remark 2 to 2.b.9]).

Setwise, the relation between the spaces is easily determined. We record it here for later use. In view of Proposition 5.2 we need only consider the case when $(m_{\nu+1} - m_\nu)_\nu$ is unbounded.

Proposition 6.5 *Let* $0 < p, q, r, s \leq \infty$, *and let* (m_ν) *be an index sequence for which* $(m_{\nu+1} - m_\nu)_\nu$ *is unbounded. Then*

$$l(\boldsymbol{m}, p, q) = l(\boldsymbol{m}, r, s) \Leftrightarrow p = q \text{ and } r = s.$$

Proposition 6.4 enables us to determine the isomorphic structure of the sequence spaces c and d. We shall employ the following terminology, see Leindler [61, 65].

A positive monotonically non-increasing (non-decreasing) sequence (c_n) is called *quasi-geometrically decreasing (increasing)* if there exists an $N \in \mathbb{N}$ such that

$$c_{n+N} \leq \frac{1}{2} c_n \quad (c_{n+N} \geq 2c_n)$$

for all n. The conditions can also be put as

$$\limsup_{n\to\infty} \frac{c_{n+N}}{c_n} < 1 \quad \left(\liminf_{n\to\infty} \frac{c_{n+N}}{c_n} > 1\right)$$

for some $N \in \mathbb{N}$. We allow the first few c_n to be infinite (zero).

Theorem 6.6 *Let* $0 < p, q \leq \infty$.
(a) *Let* $\boldsymbol{a}$ *be non-trivial, and set* $\boldsymbol{w} = (\|\boldsymbol{a} - P_n \boldsymbol{a}\|_{l^q})_n$. *Then*

$$\begin{aligned} c(\boldsymbol{a}, p, q) &\cong l(p, q) && \textit{if } \boldsymbol{w} \textit{ is not quasi-geometrically decreasing,} \\ &= l^q_{\boldsymbol{w}} \cong l^q && \textit{else.} \end{aligned}$$

(b) *Let* $\boldsymbol{a}$ *be non-trivial, and set* $\boldsymbol{w} = (\|P_n \boldsymbol{a}\|_{l^q})_n$. *Then*

$$\begin{aligned} d(\boldsymbol{a}, p, q) &\cong l(p, q) && \textit{if } \boldsymbol{w} \textit{ is not quasi-geometrically increasing,} \\ &= l^q_{\boldsymbol{w}} \cong l^q && \textit{else.} \end{aligned}$$

Proof. The result follows from Theorems 3.1, 3.2 and Proposition 6.4 once we have proved the following: If a positive monotonic sequence $\boldsymbol{s}$ with $s_n \to 0$ and an index sequence (m_ν) are correlated via (2.1), then we have

(6.1) $(m_{\nu+1} - m_\nu)_\nu$ is bounded $\Leftrightarrow$ $\boldsymbol{s}$ is quasi-geometrically decreasing.

To see this, let $m_{\nu+1} - m_\nu \leq M$ for all ν. Consider $n \in \mathbb{N}$. Then $m_\nu \leq n < m_{\nu+1}$ for some $\nu \geq 0$, so that $n + 2M \geq m_{\nu+2}$. It follows from (2.1) that

$$\frac{s_{n+2M}}{s_n} \leq \frac{1}{2}.$$

Hence $\boldsymbol{s}$ is quasi-geometrically decreasing. Conversely, if $s_{n+N} \leq s_n/2$ for all n, then $s_{m_\nu+N} \leq s_{m_\nu}/2 \leq 1/2^{\nu+1}$ for $\nu \geq 1$, so that $m_{\nu+1} \leq m_\nu + N$. Hence $(m_{\nu+1} - m_\nu)_\nu$ is bounded. □

It is easily seen that $l(p,q)$ is locally convex if and only if $p,q \geq 1$ (cf. [50], [77]). From Theorem 6.6 and Proposition 5.1 one can now decide for each $c(\boldsymbol{a},p,q)$ and $d(\boldsymbol{a},p,q)$ if it is locally convex or not. We remark that for these spaces $p,q \geq 1$ is a sufficient but not a necessary condition.

7 Multipliers and Duality

A *multiplier* from a sequence space E into a sequence space F is a sequence $\boldsymbol{y}$ such that $\boldsymbol{x} \cdot \boldsymbol{y} = (x_n y_n)_n \in F$ for every $\boldsymbol{x} \in E$. The (linear) space of all such multipliers is denoted by $\mathcal{M}(E,F)$. In particular, the *Köthe* (or α-) *dual* $E^\times$ of E is defined as the space $\mathcal{M}(E,l^1)$ of multipliers into l^1. It is well known that for every solid Fréchet K-space E containing φ with the AK-property the dual E^* is canonically isomorphic to $E^\times$ via the mapping $f \mapsto (f(\boldsymbol{e_n}))_n$, where $\boldsymbol{e_n} = (\delta_{nk})_k$ is the nth unit sequence, see [100, 7.2.9]. Hence this is true for any $c(\boldsymbol{a},p,q)$ and $d(\boldsymbol{a},p,q)$ with non-trivial $\boldsymbol{a}$ and $q < \infty$ by the previous section.

We consider here multipliers between l^p and one of the spaces c and d. Multipliers between two spaces of type c and d are more difficult to treat, unless the index sequences (m_ν) that appear after writing these spaces in block form happen to coincide. An example will be considered in Example 13.2 below.

By Observation 5.3 the present problem is reduced to characterising multipliers between $l(\boldsymbol{m},p,q)$-spaces. To state the corresponding result we define $p \to q$ for $0 < p,q \leq \infty$ by

$$\frac{1}{p \to q} = \frac{1}{q} - \frac{1}{p} \qquad \text{if} \quad q < p$$

and

$$p \to q = \infty \qquad \text{if} \quad q \geq p,$$

where $1/\infty = 0$. This notation was suggested by some authors' use of $E \to F$ for $\mathcal{M}(E,F)$.

The following result was essentially proved by Kellogg [51], see also [70, Lemma 7].

Theorem 7.1 (Kellogg) *Let* $0 < p,q,r,s \leq \infty$, *and let* $\boldsymbol{m}$ *be an index sequence. Then*

$$\mathcal{M}(l(\boldsymbol{m},p,q), l(\boldsymbol{m},r,s)) = l(\boldsymbol{m}, p \to r, q \to s).$$

With the help of Chapter I it is now an easy matter to determine multiplier spaces involving c and d. We emphasise that the conditions appearing in this section have to be interpreted in accordance with Remark 2.2(i).

Theorem 7.2 *Let $0 < p, q, r \leq \infty$, and assume that $\boldsymbol{a} \in l^q \setminus \varphi$. Then*

$$\mathcal{M}(c(\boldsymbol{a}, p, q), l^r) = d(\boldsymbol{b}, p \to r, q \to r),$$

where

$$b_n = \left(\sum_{k=n}^{\infty} a_k^q\right)^{-1/q} \qquad \text{if } q \leq r; q < \infty,$$

$$b_n = a_{n-1}^{(q-r)/r} \left(\sum_{k=n-1}^{\infty} a_k^q\right)^{(r-q)/(rq)} \left(\sum_{k=n}^{\infty} a_k^q\right)^{-1/q} \qquad \text{or}$$

$$b_n = \left[\left(\sum_{k=n}^{\infty} a_k^q\right)^{r/(r-q)} - \left(\sum_{k=n-1}^{\infty} a_k^q\right)^{r/(r-q)}\right]^{(q-r)/(rq)} \qquad \text{if } r < q < \infty,$$

$$b_n = \left(\sup_{k \geq n} a_k\right)^{-1} \qquad \text{if } r = q = \infty,$$

$$b_n = \left[\left(\sup_{k \geq n} a_k\right)^{-r} - \left(\sup_{k \geq n-1} a_k\right)^{-r}\right]^{1/r} \qquad \text{if } r < q = \infty.$$

Remark 7.3 Below we give further characterisations of $\mathcal{M}(c(\boldsymbol{a}, p, q), l^r)$ that are not in terms of the space d. They have turned out to be useful in some applications. We set $u = p \to r$, $v = q \to r$ and assume the same restrictions on $\boldsymbol{a}$ as above. Then $\boldsymbol{x} \in \mathcal{M}(c(\boldsymbol{a}, p, q), l^r)$ if and only if

$$\boldsymbol{x} \in c(\boldsymbol{b}, p \to r, q \to r)_{\boldsymbol{w}} \text{ with } b_n = a_n^{(q-r)/r}, w_n = \left(\sum_{k=n}^{\infty} a_k^q\right)^{-1/r}$$

$$\text{if } r < q < \infty,$$

$$\sum_{k=n}^{\infty} \left[\left(\sum_{j=k}^{\infty} a_j^q\right)^{-1} |x_k|\right]^u = O\left(\sum_{k=n}^{\infty} a_k^q\right)^{-u/q^*} \qquad \text{if } q \leq r < p; q < 1,$$

$$\sum_{k=1}^{n} \left[\left(\sum_{j=k}^{\infty} a_j^q\right)^{-1} |x_k|\right]^u = O\left(\sum_{k=n}^{\infty} a_k^q\right)^{-u/q^*} \qquad \text{if } 1 < q \leq r < p,$$

$$\sum_n \left(\sum_{k=n}^{\infty} a_k^q\right)^{-v/q} |x_n|^u \left(\sum_{k=n}^{\infty} |x_k|^u\right)^{v/u-1} < \infty \qquad \text{if} \quad r < p; r < q < \infty,$$

$$\sum_n \left(\sup_{k\geq n} a_k\right)^{-r} |x_n|^u \left(\sum_{k=n}^{\infty} |x_k|^u\right)^{r/u-1} < \infty \qquad \text{if} \quad r < p; q = \infty.$$

We give here the proof of these results only for one particular range of values p,q,r. All other cases as, indeed, all other proofs in this section follow in the same way and will therefore be omitted.

Proof of Theorem 7.2 and Remark 7.3 for $r < p, q < \infty$. Let the index sequence $\boldsymbol{m}$ be correlated to the sequence $\boldsymbol{s}$ with $s_n = (\sum_{k=n}^{\infty} a_k^q)^{1/q}$ in the sense of Section 2, and define the sequence $\boldsymbol{w}$ by $w_n = 1/2^\nu$ for $m_\nu \leq n < m_{\nu+1}$ $(\nu \geq 0)$. Then by Theorem 3.1(a) a sequence $\boldsymbol{x}$ belongs to $c(\boldsymbol{a},p,q)$ if and only if $(w_n x_n)_n \in l(\boldsymbol{m},p,q)$. Since $l^r = l(\boldsymbol{m},r,r)$, it follows from Kellogg's theorem that $\boldsymbol{y} \in \mathcal{M}(c(\boldsymbol{a},p,q),l^r)$ if and only if

$$\left(\frac{y_n}{w_n}\right)_n \in l(\boldsymbol{m}, p \to r, q \to r),$$

that is

$$\sum_\nu \left[2^\nu \left(\sum_{k\in I_\nu} |y_k|^u\right)^{1/u}\right]^v < \infty,$$

where we have set $u := p \to r$ and $v := q \to r$. We now apply Theorem 2.1 with $\alpha = -1, \beta = q, \gamma = -v$ and $\delta = 0$ to deduce that $\boldsymbol{y} \in \mathcal{M}(c(\boldsymbol{a},p,q),l^r)$ is characterised by

$$\sum_n \left[a_{n-1}^{q/v} \left(\sum_{k=n-1}^{\infty} a_k^q\right)^{-1/v} \left(\sum_{k=n}^{\infty} a_k^q\right)^{-1/q} \left(\sum_{k=n}^{\infty} |y_k|^u\right)^{1/u}\right]^v < \infty.$$

Alternatively, if we take $\alpha = -1, \beta = \gamma = -v$ and $\delta = 0$, then $\boldsymbol{y}$ is characterised by

$$\sum_n \left[\left(\left(\sum_{k=n}^{\infty} a_k^q\right)^{-v/q} - \left(\sum_{k=n-1}^{\infty} a_k^q\right)^{-v/q}\right)^{1/v} \left(\sum_{k=n}^{\infty} |y_k|^u\right)^{1/u}\right]^v < \infty.$$

Next, upon taking $\alpha = -1, \beta = \gamma = q$ and $\delta = -uq/r$, we see that $\boldsymbol{y}$ is characterised by

$$\sum_n \left[a_n^{q/v} \left(\sum_{k=1}^{n} \left(\big(\sum_{j=k}^{\infty} a_j^q\big)^{-1/r} |y_k| \right)^u \right)^{1/u} \right]^v < \infty.$$

And finally, if we take $\alpha = -1, \gamma = -v$ and $\delta = 0$, we deduce that $\boldsymbol{y}$ is characterised by

$$\sum_n \left(\sum_{k=n}^{\infty} a_k^q \right)^{-v/q} |y_n|^u \left(\sum_{k=n}^{\infty} |y_k|^u \right)^{v/u-1} < \infty.$$

These four conditions are the ones appearing for $r < p, q < \infty$ in Theorem 7.2 and Remark 7.3 if we note that

$$\frac{1}{v} = \frac{q-r}{rq}.$$

We add that in the case of $q = \infty$ we admit certain trivial sequences $\boldsymbol{a}$, namely $\boldsymbol{a} \in l^\infty \setminus c_0$. For these $\boldsymbol{a}$, Theorem 3.1 is not applicable but the characterising conditions extend to these $\boldsymbol{a}$ as follows immediately from Proposition 5.1. □

We illustrate these results by an Example.

Example 7.4 In connection with their work on generalisations of Hardy's inequality Johnson und Mohapatra have introduced the sequence space

$$\mathrm{ces}[p, \boldsymbol{q}] = \left\{ \boldsymbol{x} : \sum_n \left(\frac{1}{Q_n} \sum_{k=1}^{n} q_k |x_k| \right)^p < \infty \right\}.$$

Here, $0 < p \le \infty$ and $\boldsymbol{q}$ is a sequence of positive numbers with $Q_n = \sum_{k=1}^n q_k$. Their problem [46, p. 200] of determining those sequences $\boldsymbol{q}$ for which

$$\mathrm{ces}[p, \boldsymbol{q}] \supset \boldsymbol{q}^{-1} l^p (= l^p_{\boldsymbol{q}})$$

was solved by Bennett (**BIII**, p. 165). The corresponding problem of characterising the $\boldsymbol{q}$ with

$$\mathrm{ces}[p, \boldsymbol{q}] \subset \boldsymbol{q}^{-1} l^p$$

was hinted at in [46, p. 195, Remark 4]. Since this inclusion is equivalent to $\mathbf{1} \in \mathcal{M}(c(1/Q_n, 1, p), l^p)$, Theorem 7.2 shows that it is never satisfied unless the space $\mathrm{ces}[p, \boldsymbol{q}]$ becomes trivial: if $p = \infty$ and $\boldsymbol{q} \in l^1$, in which case $\mathrm{ces}[p, \boldsymbol{q}] = l^1$, or if $p < \infty$ and $\sum_n Q_n^{-p} = \infty$, in which case $\mathrm{ces}[p, \boldsymbol{q}] = \{0\}$.

As we have already explained in the Introduction, the problem of finding the dual of the Cesàro sequence space ces(p) was raised in [101] and solved by A. A. Jagers [44]. In fact, he determined the Köthe dual of any sequence space $c(\boldsymbol{a},1,p)$ for $1 \le p < \infty$ remarking that its Köthe dual is canonically isomorphic to its Banach dual (see also the beginning of this section). His solution looks rather complicated but it has the virtue of providing the dual norm. Bennett (**BIV**, 12.17) has given the much simpler representation

$$\operatorname{ces}(p)^{\times} = \left\{ \boldsymbol{x} : \sum_n \sup_{k \ge n} |x_k|^{p^*} < \infty \right\}.$$

He adds (**BIV**, p. 63) that the coincidence of this space with Jagers' $\operatorname{ces}(p)^{\times}$ is not at all obvious.

We shall here extend Bennett's result to the whole class of sequence spaces $c(\boldsymbol{a},p,q)$. In the particular case of $p=1$ our representations of the Köthe duals are much simpler than the ones obtained by Jagers, again at the price of missing the dual norm.

In the sequel let $p^{\times} = p \to 1$, that is, $p^{\times} = \infty$ if $0 < p \le 1$, $p^{\times} = (1-1/p)^{-1}$ if $1 < p < \infty$ and $p^{\times} = 1$ if $p = \infty$. Since the Köthe dual of a sequence space is defined as its set of multipliers into l^1, we obtain the following immediately from Theorem 7.2.

Corollary 7.5 *Let $0 < p,q \le \infty$, and assume that $\boldsymbol{a} \in l^q \setminus \varphi$.*
(a) *Then*

$$c(\boldsymbol{a},p,q)^{\times} = d(\boldsymbol{b},p^{\times},q^{\times}),$$

where

$$b_n = \left(\sum_{k=n}^{\infty} a_k^q \right)^{-1/q} \qquad \textit{if} \quad q \le 1,$$

$$b_n = a_{n-1}^{q-1} \left(\sum_{k=n-1}^{\infty} a_k^q \right)^{-1/q^{\times}} \left(\sum_{k=n}^{\infty} a_k^q \right)^{-1/q} \qquad \textit{or}$$

$$b_n = \left[\left(\sum_{k=n}^{\infty} a_k^q \right)^{-q^{\times}/q} - \left(\sum_{k=n-1}^{\infty} a_k^q \right)^{-q^{\times}/q} \right]^{1/q^{\times}} \qquad \textit{if} \quad 1 < q < \infty,$$

$$b_n = \left(\sup_{k \ge n} a_k \right)^{-1} - \left(\sup_{k \ge n-1} a_k \right)^{-1} \qquad \textit{if} \quad q = \infty.$$

(b) *Alternatively, if* $1 < q < \infty$, *then*

$$c(\boldsymbol{a}, p, q)^\times = c(\boldsymbol{b}, p^\times, q^\times)_{\boldsymbol{w}}$$

with

$$b_n = a_n^{q-1} \quad \textit{and} \quad w_n = \left(\sum_{k=n}^{\infty} a_k^q\right)^{-1}.$$

This corollary also solves the problem of finding the (Banach) dual of any $c(\boldsymbol{a}, p, q)$ for $q < \infty$ since it is canonically isomorphic to its Köthe dual as noted at the beginning of this section. As a consequence, representation (b) confirms that $c(\boldsymbol{a}, p, q)$ is reflexive for $1 \leq p \leq \infty, 1 < q < \infty$ and non-trivial $\boldsymbol{a}$ (see Theorem 6.2).

We next give an example that completes an investigation started in **BIV**.

Example 7.6 A major part of the investigations in **BIV** is devoted to two classes of sequences spaces, $d(\boldsymbol{a}, p)$ and $g(\boldsymbol{a}, p)$. The latter spaces are defined for $0 < p < \infty$ as

$$g(\boldsymbol{a}, p) = \left\{ \boldsymbol{x} : \sum_{k=1}^{n} |x_k|^p = O(A_n) \right\},$$

where $\boldsymbol{a}$ is a sequence of non-negative numbers with $a_1 > 0$ and $A_n = \sum_{k=1}^n a_k$. Bennett has determined the Köthe dual of $g(\boldsymbol{a}, p)$ for $1 < p < \infty$ (**BIV**, 13.14) and for $0 < p < 1$ when $\boldsymbol{a} = \mathbf{1}$ (**BIV**, 15.10). Since $g(\boldsymbol{a}, p) = c(A_n^{-1/p}, p, \infty)$, Corollary 7.5 enables us to complete this description. It shows that if $A_n \to \infty$ (the other cases being of no interest) and $p \leq 1$, we have

$$g(\boldsymbol{a}, p)^\times = \left\{ \boldsymbol{x} : \sum_n (A_n^{1/p} - A_{n-1}^{1/p}) \sup_{k \geq n} |x_k| < \infty \right\}.$$

The Köthe dual of $g(\boldsymbol{a}, 1)$ (with weights) was also calculated by Ng and Lee [79] in a slightly different form. Incidentally, the Köthe dual of $d(\boldsymbol{a}, p)$ is given in **BIV**, 12.13.

We turn to the multipliers from l^r into c.

Theorem 7.7 *Let $0 < p,q,r \le \infty$, and assume that $\boldsymbol{a} \in l^q$. Then*

$$\mathcal{M}(l^r, c(\boldsymbol{a},p,q)) = c(\boldsymbol{b}, r \to p, r \to q),$$

where

$$b_n = \left(\sum_{k=n}^{\infty} a_k^q\right)^{1/q} \qquad \textit{if} \quad r \le q < \infty,$$

$$b_n = a_n^{(r-q)/r}\left(\sum_{k=n}^{\infty} a_k^q\right)^{1/r} \qquad \textit{or}$$

$$b_n = \left[\left(\sum_{k=n}^{\infty} a_k^q\right)^{r/(r-q)} - \left(\sum_{k=n+1}^{\infty} a_k^q\right)^{r/(r-q)}\right]^{(r-q)/(qr)} \qquad \textit{if} \quad q < r < \infty,$$

$$b_n = a_n \qquad \textit{if} \quad q = \infty \quad \textit{or} \quad r = \infty.$$

Remark 7.8 We set $u = r \to p, v = r \to q$ and assume that $\boldsymbol{a} \in l^q$. Then the following conditions also characterise $\boldsymbol{x} \in \mathcal{M}(l^r, c(\boldsymbol{a},p,q))$:

$$\boldsymbol{x} \in c(\boldsymbol{b}, r \to p, r \to q)_{\boldsymbol{w}} \quad \text{with} \quad b_n = a_n^{(r-q)/r}, w_n = \left(\sum_{k=n}^{\infty} a_k^q\right)^{1/r}$$

$$\text{if} \quad q < r < \infty,$$

$$\sum_{k=1}^{n}\left[\left(\sum_{j=k}^{\infty} a_j^q\right)|x_k|\right]^u = O\left(\sum_{k=n}^{\infty} a_k^q\right)^{u/q^*} \qquad \text{if} \quad p < r \le q < 1, \boldsymbol{a} \notin \varphi,$$

$$\sum_{k=n}^{\infty}\left[\left(\sum_{j=k}^{\infty} a_j^q\right)|x_k|\right]^u = O\left(\sum_{k=n}^{\infty} a_k^q\right)^{u/q^*} \qquad \text{if} \quad p < r \le q < \infty, q > 1,$$

$$\sum_{k=1}^{n} a_k^q \left(\sum_{j=1}^{k} |x_j|^u\right)^{(\sigma+q)/u} = O\left(\sum_{k=1}^{n} |x_k|^u\right)^{\sigma/u} \quad \text{for all/some} \quad \sigma > 0$$

$$\text{if} \quad p < r \le q < \infty,$$

$$\sum_n \left(\sum_{k=n}^{\infty} a_k^q\right)^{v/q} |x_n|^u \left(\sum_{k=1}^{n} |x_k|^u\right)^{v/u-1} < \infty \qquad \text{if} \quad p,q < r < \infty.$$

Next we consider multipliers involving the space d. The additional assumption $a_1 > 0$ is made in order to avoid division by 0.

Theorem 7.9 *Let $0 < p, q \leq \infty$, and assume that $\boldsymbol{a} \notin l^q$ and $a_1 > 0$. Then*

$$\mathcal{M}(d(\boldsymbol{a}, p, q), l^r) = c(\boldsymbol{b}, p \to r, q \to r),$$

where

$$b_n = \left(\sum_{k=1}^n a_k^q\right)^{-1/q} \qquad \textit{if} \quad q \leq r; q < \infty,$$

$$b_n = a_{n+1}^{(q-r)/r} \left(\sum_{k=1}^{n+1} a_k^q\right)^{(r-q)/(rq)} \left(\sum_{k=1}^n a_k^q\right)^{-1/q} \qquad \textit{or}$$

$$b_n = \left[\left(\sum_{k=1}^n a_k^q\right)^{r/(r-q)} - \left(\sum_{k=1}^{n+1} a_k^q\right)^{r/(r-q)}\right]^{(q-r)/(rq)} \qquad \textit{if} \quad r < q < \infty,$$

$$b_n = \left(\sup_{k \leq n} a_k\right)^{-1} \qquad \textit{if} \quad r = q = \infty,$$

$$b_n = \left[\left(\sup_{k \leq n} a_k\right)^{-r} - \left(\sup_{k \leq n+1} a_k\right)^{-r}\right]^{1/r} \qquad \textit{if} \quad r < q = \infty.$$

Remark 7.10 We set $u = p \to r, v = q \to r$ and assume (only) that $a_1 > 0$. Then the following conditions that are not in terms of the spaces c also characterise when $\boldsymbol{x} \in \mathcal{M}(d(\boldsymbol{a}, p, q), l^r)$:

$$\boldsymbol{x} \in d(\boldsymbol{b}, r \to p, r \to q)_{\boldsymbol{w}} \quad \text{with} \quad b_n = a_n^{(q-r)/r}, w_n = \left(\sum_{k=1}^n a_k^q\right)^{-1/r} \qquad \text{if} \quad r < q < \infty,$$

$$\sum_{k=1}^n \left[\left(\sum_{j=1}^k a_j^q\right)^{-1} |x_k|\right]^u = O\left(\sum_{k=1}^n a_k^q\right)^{-u/q^*} \qquad \text{if} \quad q \leq r < p; q < 1,$$

$$\sum_{k=n}^\infty \left[\left(\sum_{j=1}^k a_j^q\right)^{-1} |x_k|\right]^u = O\left(\sum_{k=1}^n a_k^q\right)^{-u/q^*} \qquad \text{if} \quad 1 < q \leq r < p,$$

$$\sum_n \left(\sum_{k=1}^n a_k^q\right)^{-v/q} |x_n|^u \left(\sum_{k=1}^n |x_k|^u\right)^{v/u - 1} < \infty \qquad \text{if} \quad r < p; r < q < \infty,$$

$$\sum_n \left(\sup_{k \leq n} a_k\right)^{-r} |x_n|^u \left(\sum_{k=1}^n |x_k|^u\right)^{r/u - 1} < \infty \qquad \text{if} \quad r < p; q = \infty.$$

In the particular case of $r = 1$, Theorem 7.9 leads to the determination of the Köthe duals of $d(\boldsymbol{a},p,q)$ and hence, for $q < \infty$, their (Banach) duals. Compare the discussion in connection with Corollary 7.5.

Corollary 7.11 *Let* $0 < p, q \leq \infty$.
(a) *Assume that* $\boldsymbol{a} \notin l^q$ *and* $a_1 > 0$. *Then*

$$d(\boldsymbol{a},p,q)^\times = c(\boldsymbol{b},p^\times,q^\times),$$

where

$$b_n = \left(\sum_{k=1}^n a_k^q\right)^{-1/q} \qquad \textit{if} \quad q \leq 1,$$

$$b_n = a_{n+1}^{q-1}\left(\sum_{k=1}^{n+1} a_k^q\right)^{-1/q^\times}\left(\sum_{k=1}^{n} a_k^q\right)^{-1/q} \qquad \textit{or}$$

$$b_n = \left[\left(\sum_{k=1}^n a_k^q\right)^{-q^\times/q} - \left(\sum_{k=1}^{n+1} a_k^q\right)^{-q^\times/q}\right]^{1/q^\times} \qquad \textit{if} \quad 1 < q < \infty,$$

$$b_n = \left(\sup_{k\leq n} a_k\right)^{-1} - \left(\sup_{k\leq n+1} a_k\right)^{-1} \qquad \textit{if} \quad q = \infty.$$

(b) *Alternatively, if* $1 < q < \infty$ *and* $a_1 > 0$, *then*

$$d(\boldsymbol{a},p,q)^\times = d(\boldsymbol{b},p^\times,q^\times)_{\boldsymbol{w}}$$

with

$$b_n = a_n^{q-1} \quad \textit{and} \quad w_n = \left(\sum_{k=1}^n a_k^q\right)^{-1}.$$

This result generalises **BIV**, Corollary 12.13 ($p = \infty, q < \infty$) and Theorem 16.2 ($p = 1, q < \infty$). We consider two more examples from that paper.

Example 7.12 (a) For $1 < p < \infty$, Zhang (see [56], [91]) has introduced the *reverse Cesàro sequence space*

$$\mathrm{zha}(p) = \left\{\boldsymbol{x} : \sum_n \left(\frac{1}{n} - \frac{1}{n+1}\right)\left(n\sum_{k=n}^{\infty}|x_k|\right)^p < \infty\right\}$$

(Bennett's notation). This space coincides with $d(n^{1-2/p}, 1, p)$. Hence we obtain

$$\text{zha}(p)^\times = \left\{ x : \sum_n \frac{1}{n^2} \sup_{k \le n} |x_k|^{p^\times} < \infty \right\}.$$

(One might replace $1/n^2$ here by $1/n - 1/(n+1)$ to increase the symmetry). This description is simpler than the one given in **BIV**, p. 85.

(b) In **BIV**, §18, Bennett studies the behaviour of the weighted means

$$\mathfrak{M}_r(y) = \left(\left(\frac{\sum_{k=1}^n a_k |y_k|^r}{\sum_{k=1}^n a_k} \right)^{1/r} \right)_n,$$

where a is a fixed sequence of non-negative terms with $a_1 > 0$ and r is real parameter (cf. [40, 2.2]). In Theorem 18.12 of **BIV** he characterises without proof when a sequence x satisfies, coordinatewise,

$$|x| \le \mathfrak{M}_r(y)$$

for some $y \in l^p$, where $p > 0$ and $r < p$ are fixed. Now if $r > 0$, a proof of this characterisation can be obtained directly with part (b) of the Corollary. One need only realise that by Theorem 17.20 of **BIV**, $|x| \le \mathfrak{M}_r(y)$ for some $y \in l^p$ is equivalent to $(A_n |x_n|^r)_n \in d(a, 1, (p/r)^\times)^\times$.

Theorem 7.13 *Let $0 < p, q \le \infty$. Then*

$$\mathcal{M}(l^r, d(a, p, q)) = d(b, r \to p, r \to q),$$

where

$$b_n = \left(\sum_{k=1}^n a_k^q \right)^{1/q} \qquad \textit{if} \quad r \le q < \infty,$$

$$b_n = a_n^{(r-q)/r} \left(\sum_{k=1}^n a_k^q \right)^{1/r} \qquad \textit{or}$$

$$b_n = \left[\left(\sum_{k=1}^n a_k^q \right)^{r/(r-q)} - \left(\sum_{k=1}^{n-1} a_k^q \right)^{r/(r-q)} \right]^{(r-q)/(qr)} \qquad \textit{if} \quad q < r < \infty,$$

$$b_n = a_n \qquad \textit{if} \quad q = \infty \quad \textit{or} \quad r = \infty.$$

Remark 7.14 We set $u = r \to p$ and $v = r \to q$. Then the following conditions also characterise $\boldsymbol{x} \in \mathcal{M}(l^r, d(\boldsymbol{a},p,q))$:

$$\boldsymbol{x} \in d(\boldsymbol{b}, r \to p, r \to q)_{\boldsymbol{w}} \quad \text{with} \quad b_n = a_n^{(r-q)/r}, w_n = \left(\sum_{k=1}^{n} a_k^q\right)^{1/r}$$

if $q < r < \infty$,

$$\sum_{k=n}^{\infty}\left[\left(\sum_{j=1}^{k} a_j^q\right)|x_k|\right]^u = O\left(\sum_{k=1}^{n} a_k^q\right)^{u/q^*} \qquad \text{if } p < r \le q < 1, a_1 > 0,$$

$$\sum_{k=1}^{n}\left[\left(\sum_{j=1}^{k} a_j^q\right)|x_k|\right]^u = O\left(\sum_{k=1}^{n} a_k^q\right)^{u/q^*} \qquad \text{if } p < r \le q < \infty, q > 1,$$

$$\sum_{k=n}^{\infty} a_k^q\left(\sum_{j=k}^{\infty}|x_j|^u\right)^{(\sigma+q)/u} = O\left(\sum_{k=n}^{\infty}|x_k|^u\right)^{\sigma/u} \quad \text{for all/some } \sigma > 0$$

if $p < r \le q < \infty$,

$$\sum_n\left(\sum_{k=1}^{n} a_k^q\right)^{v/q}|x_n|^u\left(\sum_{k=n}^{\infty}|x_k|^u\right)^{v/u-1} < \infty \qquad \text{if } p,q < r < \infty.$$

It is remarkable that the last two results hold essentially for all $\boldsymbol{a}$.

8 Products and Factorisations

The major new concept introduced by Bennett in **BIV** is that of factorising an inequality. Let us take Hardy's inequality again as paradigm to illustrate this concept. As we have seen in the Introduction, this inequality is equivalent, in its qualitative version, to the inclusion

$$l^p \subset \text{ces}(p). \tag{8.1}$$

This can also be interpreted as saying that the sequence $\mathbf{1} = (1,1,1,\ldots)$ is a multiplier from l^p to $\text{ces}(p)$. One might wonder how much room there is between l^p and $\text{ces}(p)$. By the definition of multipliers it is clear that

$$l^p \cdot \mathcal{M}(l^p, \text{ces}(p)) \subset \text{ces}(p), \tag{8.2}$$

where the product of two sequence spaces E and F is defined as $E \cdot F = \{\boldsymbol{x}\cdot\boldsymbol{y} = (x_n y_n)_n : \boldsymbol{x} \in E, \boldsymbol{y} \in F\}$, and $\mathcal{M}$ cannot be replaced by any larger space. What is surprising now is that (8.2) is in fact an identity. Since $\mathcal{M}(l^p, \text{ces}(p)) = \{\boldsymbol{x} : \sup_n \frac{1}{n}\sum_{k=1}^n |x_k|^{p^*} < \infty\} =: g(p^*)$ (**BIV**, cf. Theorem 7.7 above), (8.2) then becomes

$$l^p \cdot g(p^*) = \text{ces}(p). \tag{8.3}$$

Since clearly $\mathbf{1} \in g(p^*)$, Hardy's inequality (8.1) is thus replaced by an equality which Bennett calls its enhanced version. We add that there is also a quantitative extension of (8.3) that enhances in turn the quantitative version of Hardy's inequality, compare Theorem 4.5 of **BIV**.

In another part of his work Bennett searches for representations of products like

$$l^p \cdot \left\{ \boldsymbol{x} : \sup_n a_n \sum_{k=n}^{\infty} |x_k|^q < \infty \right\},$$

cf. Section 3 in **BIII** and §5 in **BIV**. What is surprising here is that this product can be calculated and can be described in rather simple terms.

These two examples may serve as motivation for the type of problems that we want to study in this section. Given two sequence spaces E and F of type l^p, c or d, the task is to find a representation of the product

$$E \cdot F$$

and to derive a factorisation

$$E = F \cdot G$$

with a suitable G, if it exists.

Having the transformation rules of Chapter I at our disposal we can deduce such results immediately from product and factorisation results for $l(p,q)$-spaces. The latter are merely extensions of Hölder's inequality and its converse. We shall in the sequel employ the following notation. If $0 < p, q \leq \infty$, then $p * q$ is defined by

$$\frac{1}{p*q} = \frac{1}{p} + \frac{1}{q},$$

with the usual interpretation if p or q is infinite. This notation was suggested by some authors' use of $E * F$ for $E \cdot F$. We remark that

$$p * (p \to q) = \min(p, q) \quad \text{and} \quad p \to (p * q) = q$$

for any p and q, two formulae that will tacitly be used throughout this section.

Theorem 8.1 *Let $0 < p,q,r,s \le \infty$, and let $\boldsymbol{m}$ be an index sequence. Then*

$$l(\boldsymbol{m},p,q)\cdot l(\boldsymbol{m},r,s) = l(\boldsymbol{m},p*r,q*s).$$

Proof. Since $E\cdot\mathcal{M}(E,F)\subset F$ holds for all sequence spaces E and F, one half of the theorem follows directly from Kellogg's theorem 7.1. It remains to show that every $\boldsymbol{x}\in l(\boldsymbol{m},p*r,q*s)$ admits a factorisation $\boldsymbol{y}\cdot\boldsymbol{z}$ with $\boldsymbol{y}\in l(\boldsymbol{m},p,q)$ and $\boldsymbol{z}\in l(\boldsymbol{m},r,s)$, where we may assume that all $x_n>0$. If all parameters are finite, one can take $y_n = x_n^{\alpha}\rho_\nu^{\beta}$ and $z_n = x_n^{\gamma}\rho_\nu^{\delta}$ for $m_\nu\le n< m_{\nu+1}, \nu\ge 0$, where $\alpha=(p*r)/p, \beta=(q*s)/q-(p*r)/p, \gamma=(p*r)/r, \delta=(q*s)/s-(p*r)/r$ and $\rho_\nu = \left(\sum_{k=m_\nu}^{m_{\nu+1}-1} x_k^{p*r}\right)^{1/(p*r)}$. The remaining parameters are treated similarly. □

In [24, Theorems 9 and 10], Buntinas has obtained two closely related results. In fact the theorem implies his results if $p*r\ge 1$ and $q*s\ge 1$.

Theorem 8.1 enables us to prove the corresponding result if a space of type c or d and a space l^p is involved. As in Section 7 one encounters a problem, however, when trying to calculate the product of two spaces of type c and d because they usually translate into $l(p,q)$-spaces with different blocks.

The proofs in this section are completely analogous to those in Section 7 and are therefore omitted. We refer again to Remark 2.2(i) for a proper interpretation of the conditions encountered.

Theorem 8.2 *Let $0 < p,q \le \infty$.*

(a) *Assume that $\boldsymbol{a}\in l^q$. Then*

$$c(\boldsymbol{a},p,q)\cdot l^r = c(\boldsymbol{b},p*r,q*r),$$

where

$$b_n = a_n^{(q+r)/r}\left(\sum_{k=n}^{\infty} a_k^q\right)^{-1/r} \qquad \textit{or}$$

$$b_n = \left[\left(\sum_{k=n}^{\infty} a_k^q\right)^{r/(q+r)} - \left(\sum_{k=n+1}^{\infty} a_k^q\right)^{r/(q+r)}\right]^{(q+r)/(qr)} \qquad \textit{if}\quad q,r<\infty,$$

$$b_n = \left[\left(\sup_{k\ge n} a_k\right)^r - \left(\sup_{k\ge n+1} a_k\right)^r\right]^{1/r} \qquad \textit{if}\quad r<q=\infty,$$

$$b_n = a_n \qquad \textit{if}\quad r=\infty.$$

(b) *Assume that* $a_1 > 0$. *Then*

$$d(\boldsymbol{a}, p, q) \cdot l^r = d(\boldsymbol{b}, p * r, q * r),$$

where

$$b_n = a_n^{(q+r)/r} \left(\sum_{k=1}^{n} a_k^q \right)^{-1/r} \qquad \textit{or}$$

$$b_n = \left[\left(\sum_{k=1}^{n} a_k^q \right)^{r/(q+r)} - \left(\sum_{k=1}^{n-1} a_k^q \right)^{r/(q+r)} \right]^{(q+r)/(qr)} \qquad \textit{if} \quad q, r < \infty,$$

$$b_n = \left[\left(\sup_{k \le n} a_k \right)^r - \left(\sup_{k \le n-1} a_k \right)^r \right]^{1/r} \qquad \textit{if} \quad r < q = \infty,$$

$$b_n = a_n \qquad \textit{if} \quad r = \infty.$$

Remark 8.3 There are further characterisations of the products considered above that are useful in applications. We set $u = p * r$ and $v = q * r$. Then under the same assumptions on $\boldsymbol{a}$ as above we have:
(a) $\boldsymbol{x} \in c(\boldsymbol{a}, p, q) \cdot l^r$ if and only if

$$\boldsymbol{x} \in c(\boldsymbol{b}, p * r, q * r)_{\boldsymbol{w}} \quad \text{with} \quad b_n = a_n^{(q+r)/r}, w_n = \left(\sum_{k=n}^{\infty} a_k^q \right)^{-1/r}$$

$$\text{if} \quad q, r < \infty, \boldsymbol{a} \notin \varphi,$$

$$\sum_n \left(\sum_{k=n}^{\infty} a_k^q \right)^{v/q} |x_n|^u \left(\sum_{k=1}^{n} |x_k|^u \right)^{v/u - 1} < \infty \qquad \text{if} \quad q, r < \infty,$$

$$\sum_n \left(\sup_{k \ge n} a_k \right)^r |x_n|^u \left(\sum_{k=1}^{n} |x_k|^u \right)^{r/u - 1} < \infty \qquad \text{if} \quad r < q = \infty.$$

(b) $\boldsymbol{x} \in d(\boldsymbol{a}, p, q) \cdot l^r$ if and only if

$$\boldsymbol{x} \in d(\boldsymbol{b}, p * r, q * r)_{\boldsymbol{w}} \text{ with } b_n = a_n^{(q+r)/r}, w_n = \left(\sum_{k=1}^{n} a_k^q \right)^{-1/r}$$

$$\text{if} \quad q, r < \infty, a_1 > 0,$$

$$\sum_n \left(\sum_{k=1}^{n} a_k^q \right)^{v/q} |x_n|^u \left(\sum_{k=n}^{\infty} |x_k|^u \right)^{v/u - 1} < \infty \qquad \text{if} \quad q, r < \infty,$$

$$\sum_n \left(\sup_{k \le n} a_k \right)^r |x_k|^u \left(\sum_{k=n}^{\infty} |x_k|^u \right)^{r/u - 1} < \infty \qquad \text{if} \quad r < q = \infty.$$

Example 8.4 Let $1 < p < \infty$ and $0 < q < \infty$. Then

$$\operatorname{ces}(p) \cdot l^q = \left\{ \boldsymbol{x} : \sum_n \left(\frac{1}{n} \sum_{k=1}^{n} |x_k|^{q/(q+1)} \right)^{p(q+1)/(q+p)} < \infty \right\}.$$

It is remarkable that this product space is never locally convex (under its natural topology) whatever p and q are, as follows from the last remark in Section 6.

We turn to the related problem of factorising a given space. Again we first need the pertinent result for $l(p, q)$-spaces.

Theorem 8.5 *Let* $0 < p, q, r, s \leq \infty$. *If* $\boldsymbol{m}$ *is an index sequence such that* $(m_{\nu+1} - m_\nu)_\nu$ *is unbounded* [$(m_{\nu+1} - m_\nu)_\nu$ *is bounded*], *then there is a sequence space* G *with*

$$l(\boldsymbol{m}, r, s) = l(\boldsymbol{m}, p, q) \cdot G$$

if and only if $r \leq p$ *and* $s \leq q$ [*if and only if* $s \leq q$]. *In that case* G *can be taken as*

$$G = l(\boldsymbol{m}, p \to r, q \to s).$$

Proof. Let G be a sequence space with $l(\boldsymbol{m}, r, s) = l(\boldsymbol{m}, p, q) \cdot G$. Then G is contained in $\mathcal{M}(l(\boldsymbol{m}, p, q), l(\boldsymbol{m}, r, s))$, which is $l(\boldsymbol{m}, p \to r, q \to s)$ by Theorem 7.1. With this we can deduce, using Theorem 8.1, that

$$\begin{aligned} l(\boldsymbol{m}, r, s) \subset l(\boldsymbol{m}, p, q) \cdot l(\boldsymbol{m}, p \to r, q \to s) &= l(\boldsymbol{m}, p * (p \to r), q * (q \to s)) \\ &= l(\boldsymbol{m}, \min(p, r), \min(q, s)) \subset l(\boldsymbol{m}, r, s), \end{aligned}$$

hence that we have equality throughout. Now, Propositions 6.4 and 5.2 imply the necessity of the stated condition. Its sufficiency follows readily from Theorem 8.1 and Propositions 6.4 and 5.2. □

We shall employ this result to study factorisations involving the spaces c and d, concentrating on those $\boldsymbol{a}$ for which c or d do not reduce to "trivial" spaces like l^p. The multiplier spaces that will appear can be looked up in Section 7.

Theorem 8.6 *Let* $0 < p, q, r \leq \infty$.

(a) *Let* $\boldsymbol{a} \in l^q \setminus \varphi$ *if* $q < \infty$, $\boldsymbol{a} \in c_0 \setminus \varphi$ *if* $q = \infty$ *and assume that* $(\|\boldsymbol{a} - P_n\boldsymbol{a}\|_q)_n$ *is not quasi-geometrically decreasing. Then:*

(i) *There is a sequence space* G *with*

$$c(\boldsymbol{a}, p, q) = l^r \cdot G$$

if and only if $p \leq r$ *and* $q \leq r$. *In that case,* G *can be taken as* $\mathcal{M}(l^r, c(\boldsymbol{a}, p, q))$.

(ii) *There is a sequence space* G *with*

$$l^r = c(\boldsymbol{a}, p, q) \cdot G$$

if and only if $r \leq p$ *and* $r \leq q$. *In that case,* G *can be taken as* $\mathcal{M}(c(\boldsymbol{a}, p, q), l^r)$.

(b) *The same results hold for* $d(\boldsymbol{a}, p, q)$ *if* $\boldsymbol{a} \notin l^q$ *and* $(\|P_n\boldsymbol{a}\|_q)_n$ *is not quasi-geometrically increasing.*

Proof. This follows from Theorems 3.1, 3.2 and 8.5 together with the equivalence (6.1) in the proof of Theorem 6.6. The simple details are omitted. □

Example 8.7 (Cf. **BIV**, 1.5 and §15). If $1 < p < \infty$ and $0 < q \leq \infty$, then

$$\text{(a)}\quad \mathrm{ces}(p) = l^q \cdot \left\{ \boldsymbol{x} : \sum_n \left(\frac{1}{n} \sum_{k=1}^n |x_k|^{q^*} \right)^{(q \to p)/q^*} < \infty \right\} \quad \text{if } q > p,$$

$$= l^q \cdot \left\{ \boldsymbol{x} : \sup_n \left(\frac{1}{n} \sum_{k=1}^n |x_k|^{p^*} \right) < \infty \right\} \quad \text{if } q = p,$$

$$\neq l^q \cdot G \quad \text{for any sequence space } G \quad \text{if } q < p;$$

$$\text{(b)}\quad l^q = \mathrm{ces}(p) \cdot \left\{ \boldsymbol{x} : \sum_n \left(\frac{1}{n} \sum_{k=n}^\infty |x_k|^{-q^*} \right)^{-(p \to q)/q^*} < \infty \right\} \quad \text{if } q < 1,$$

$$= \mathrm{ces}(p) \cdot \left\{ \boldsymbol{x} : \sum_n \sup_{k \geq n} |x_k|^{p^*} < \infty \right\} \quad \text{if } q = 1,$$

$$\neq \mathrm{ces}(p) \cdot G \quad \text{for any sequence space } G \quad \text{if } q > 1.$$

In connection with the determination of various multiplier spaces Bennett (**BIV**, §14) has introduced an interesting new concept, the cancellation property. A sequence space F is said to have the *G-cancellation property* if the inclusion $\boldsymbol{x} \cdot G \subset F \cdot G$ implies that $\boldsymbol{x}$ belongs to F. Equivalent statements are

$$\mathcal{M}(G, F \cdot G) = F$$

and

$$\mathcal{M}(E, F) = \mathcal{M}(E \cdot G, F \cdot G) \quad \text{for any sequence space} \quad E,$$

from which the usefulness of this concept in the study of multiplier spaces is evident – provided that interesting spaces with the cancellation property exist. Now, Bennett has verified that all of the main spaces considered by him in **BIV** have the cancellation property with respect to each other (Proposition 14.5). We shall yet again show that this is true because it is true for $l(p, q)$-spaces. We start by extending **BIV**, Lemma 14.6.

Lemma 8.8 *If a sequence space E has the l^p-cancellation property for every $p > 0$, then it has the $l(\boldsymbol{m}, p, q)$-cancellation property for every $p, q > 0$ and every $\boldsymbol{m}$.*

Proof. Let $\boldsymbol{x}$ satisfy $\boldsymbol{x} \cdot l(\boldsymbol{m}, p, q) \subset E \cdot l(\boldsymbol{m}, p, q)$. If $r = \min(p, q)$, then by Theorem 8.5 there is a sequence space G with $l^r = l(\boldsymbol{m}, p, q) \cdot G$. It follows that

$$\begin{aligned} \boldsymbol{x} \cdot l^r &= \boldsymbol{x} \cdot l(\boldsymbol{m}, p, q) \cdot G \\ &\subset E \cdot l(\boldsymbol{m}, p, q) \cdot G \\ &= E \cdot l^r. \end{aligned}$$

Since E has the l^r-cancellation property, we deduce that $\boldsymbol{x} \in E$, which had to be shown. □

It is a direct consequence of Theorems 7.1 and 8.1 that $l(\boldsymbol{m}, p, q)$ has the l^r-cancellation property for any $p, q, r > 0$. The lemma thus implies:

Theorem 8.9 *If $0 < p, q, r, s \leq \infty$, then $l(\boldsymbol{m}, p, q)$ has the $l(\boldsymbol{n}, r, s)$-cancellation property for arbitrary index sequences $\boldsymbol{m}$ and $\boldsymbol{n}$.*

This result is remarkable in that it is the only one in the present notes that allows two $l(p,q)$-spaces with different blocks.

From Theorem 8.9 we immediately obtain cancellation properties for the spaces c and d. We use the fact that the cancellation property is preserved if weights are introduced in the two spaces involved.

Theorem 8.10 *For* $0 < p,q,r,s \leq \infty$, *the spaces* $c(\boldsymbol{a},p,q)$, $d(\boldsymbol{a},p,q)$ *and* l^p *have the* $c(\boldsymbol{b},r,s)$-, $d(\boldsymbol{b},r,s)$- *and* l^r*-cancellation properties, provided that* $c(\boldsymbol{b},r,s) \neq \{0\},\omega$.

With this extension of **BIV**, Proposition 14.5 we end our study of the sequence spaces c and d for now. Two more results will be added in Section 13. We remark that with few exceptions like Theorem 6.9 in **BIV** all of Bennett's results concerning these spaces have here been re-proved and many of them extended, in so far as their qualitative content is concerned (see our comments in the Introduction).

We next turn to applications of the results of this chapter.

Chapter III

Applications to Matrix Operators and Inequalities

9 Factorable Matrices as Operators on l^p and Hardy's Inequality

In some sense we now come to the central section of these notes. We shall here confirm the conjecture of Bennett that motivated our research.

In order to describe the contents of this conjecture we take another look at Hardy's inequality. Since l^p is a solid sequence space, it is equivalent to the inequality

$$\left(\sum_n \left|\frac{1}{n}\sum_{k=1}^n x_k\right|^p\right)^{1/p} \leq K\left(\sum_n |x_n|^p\right)^{1/p}.$$

In view of the closed graph theorem this in turn is equivalent to the assertion that the Cesàro operator $C_1 : \boldsymbol{x} \mapsto \left(\frac{1}{n}\sum_{k=1}^n x_k\right)_n$ maps l^p into itself. C_1 can be interpreted as a multiplication of the Cesàro matrix $C_1 = (\frac{1}{n})_{n\geq 1, k\leq n}$ with $\boldsymbol{x}$. In this way Hardy's inequality links up with the so-called matrix mapping problem which consists in characterising those matrices $(a_{nk})_{n,k\geq 1}$ that map l^p into l^q. Since this problem is open for most p and q (in particular if $1 < p, q < \infty$ unless $p = q = 2$, see [89]), one is interested in finding a solution for subclasses of matrices; we refer to Section 8 of **BI** for a good discussion. The class of matrices considered by Bennett consists of the *factorable matrices* $M(\boldsymbol{a}, \boldsymbol{b})$ for non-negative sequences $\boldsymbol{a}$ and $\boldsymbol{b}$; their entries are defined by

$$a_{nk} = \begin{cases} a_n b_k & \text{if } k \leq n \\ 0 & \text{if } k > n, \end{cases}$$

cf. [98, p. 649]. This class of matrices contains the (positive) weighted mean matrices which constitute, alongside the Nörlund and Hausdorff matrices, one of the most important classes of matrices in classical summability theory. They are exactly the factorable matrices with row sums 1 (that is, with $b_1 > 0$ and $a_n = (\sum_{k=1}^{n} b_k)^{-1}$ for $n \in \mathbb{N}$). In that sense, the factorable matrices could also be called *generalised weighted mean matrices*. Of course, C_1 is the prime example.

In **BI** and **BIII**, Bennett has completely solved the l^p-l^q-mapping problem for the class of factorable matrices. Equivalently, he has characterised those sequences $\boldsymbol{a}$ and $\boldsymbol{b}$ for which Hardy's inequality with weights

$$\left(\sum_n \left(a_n \sum_{k=1}^{n} b_k x_k\right)^q\right)^{1/q} \leq K \left(\sum_n x_n^p\right)^{1/p} \tag{9.1}$$

holds with some $K > 0$ for all non-negative sequences $\boldsymbol{x}$. The result breaks up into several cases depending on p and q, see **BIII**, Theorem 1. Bennett notes that his condition for the case $0 < q < p < 1$ has a more complicated structure than the others, and he asks (**BIII**, pp. 160-161) if it can be replaced by another, simpler condition, (39) of **BIII**. We shall here confirm this. In fact, our work in Chapter II allows us to write down immediately the characterising conditions for the whole range of p and q, thus providing a new and elementary proof of Bennett's results. This is due to the following simple observation.

Observation 9.1 *Let $0 < p, q \leq \infty$. Then a factorable matrix $M(\boldsymbol{a}, \boldsymbol{b})$ maps l^p into l^q if and only if $\boldsymbol{b}$ is a multiplier from l^p into $c(\boldsymbol{a}, 1, q)$.*

With this, the characterising conditions can be read off from Theorem 7.7 and Remark 7.8 (the restriction $\boldsymbol{a} \in l^q$ is easily dealt with). In particular we can give the announced positive answer to Bennett's question.

Theorem 9.2 *Let $0 < q < p \le 1$. Then a factorable matrix $M(\boldsymbol{a}, \boldsymbol{b})$ maps l^p into l^q if and only if*

$$\sum_n a_n^q \left(\sum_{k=n}^{\infty} a_k^q\right)^{r/p} \sup_{k\le n} b_k^r < \infty, \tag{9.2ix}$$

where $\frac{1}{r} = \frac{1}{q} - \frac{1}{p}$.

The sufficieny of this condition was already noted by Bennett. A stronger sufficient condition was also given recently by Leindler [62, Theorem 1']. We shall comment on Leindler's paper at the end of Section 10.

We add a remark on the corresponding condition for the case $0 < q < p, 1 < p < \infty$. Bennett (**BIII**, (32)) provides

$$\sum_n \left(\sum_{k=n}^{\infty} a_k^q\right)^{r/q} b_n^{p^*} \left(\sum_{k=1}^{n} b_k^{p^*}\right)^{r/q^*} < \infty \tag{9.2x}$$

as an alternative condition, with r as in the theorem, but he states that his proof requires that in addition $q \ge 1$. On the other hand, essentially the same problem (a simple change of variables is required) was investigated by Braverman and Stepanov [22] who were apparently unaware of Bennett's work. They also come up with condition (9.2x), but they only show the equivalence for $q < 1$. Our approach confirms (9.2x) as a characterising condition with no additional restriction on q (see the last case in Remark 7.8).

Now that we know when $M(\boldsymbol{a}, \boldsymbol{b})$ maps l^p into l^q one may ask when it even represents a compact operator. It follows from a result of Brown, Halmos and Shields [23] that C_1 is not a compact operator on l^2. Indeed, it is not compact on any $l^p, p > 1$. More generally, Rhaly [84] and Leibowitz [57] have investigated the compactness of the factorable matrices $R_{\boldsymbol{a}} = M(\boldsymbol{a}, \mathbf{1})$, sometimes called *Rhaly matrices*, and have obtained some partial results. The complete picture will be a consequence of the following theorem that extends some results of Johnson and Mohapatra [45]. We first explain the setting. If $A = (a_{nk})$ is a matrix with non-negative entries, we set

$$A(p) = \left\{ \boldsymbol{x} : \sum_n \left(\sum_k a_{nk}|x_k|\right)^p < \infty \right\}$$

for $0 < p \le \infty$, with the usual modification if $p = \infty$. This space is $A(1,p)$ of Remark 6.3, see also **BIV**, §17. We assume that *no column of A vanishes.*

Then $A(p)$ becomes a quasi-Banach space under the obvious quasi-norm. For a sequence $\boldsymbol{b}$ we denote by $D_{\boldsymbol{b}}$ the diagonal matrix with diagonal $\boldsymbol{b}$. Then every multiplier $\boldsymbol{b} \in \mathcal{M}(l^p, A(q))$ induces a matrix mapping $AD_{\boldsymbol{b}} : l^p \to l^q$, for $0 < p, q \leq \infty$.

Theorem 9.3 *Let A be row-finite and $\boldsymbol{b} \in \mathcal{M}(l^p, A(q))$. Then the mapping $AD_{\boldsymbol{b}} : l^p \to l^q$ is compact if and only if $P_n\boldsymbol{b} \to \boldsymbol{b}$ in $\mathcal{M}(l^p, A(q))$ with respect to the operator topology.*

Proof. If $P_n\boldsymbol{b} \to \boldsymbol{b}$, then $D_{\boldsymbol{b}} : l^p \to A(q)$ is compact as limit of finite rank operators. The continuity of $A : A(q) \to l^q$ then implies the compactness of $AD_{\boldsymbol{b}}$.

Conversely assume that $AD_{\boldsymbol{b}}$ is compact. Since

$$\|\boldsymbol{b} - P_n\boldsymbol{b}\|_{\mathcal{M}(l^p, A(q))} = \sup \left\{\|AD_{\boldsymbol{b}}(\boldsymbol{x} - P_n\boldsymbol{x})\|_{l^q} : \|\boldsymbol{x}\|_{l^p} \leq 1\right\},$$

we have to show that the right hand side tends to 0 as $n \to \infty$. Else we can find a $\delta > 0$ and $\boldsymbol{x}^{(n)} \in l^p$ with $\|\boldsymbol{x}^{(n)}\| \leq 1$ and $m_n \geq n$ such that

$$\|AD_{\boldsymbol{b}}(\boldsymbol{x}^{(n)} - P_{m_n}\boldsymbol{x}^{(n)})\|_{l^q} \geq \delta \tag{9.3}$$

for all n. We set $\boldsymbol{y}^{(n)} = \boldsymbol{x}^{(n)} - P_{m_n}\boldsymbol{x}^{(n)}$. Then by compactness of $AD_{\boldsymbol{b}}$ we can assume, after passing to a subsequence, that $AD_{\boldsymbol{b}}\boldsymbol{y}^{(n)}$ converges in l^q. Since A is row-finite and $\boldsymbol{y}^{(n)} \to \boldsymbol{0}$ coordinatewise, we see that $AD_{\boldsymbol{b}}\boldsymbol{y}^{(n)} \to \boldsymbol{0}$ coordinatewise, hence that $AD_{\boldsymbol{b}}\boldsymbol{y}^{(n)} \to \boldsymbol{0}$ in l^q. This contradiction to (9.3) completes the proof. □

This result is due to Johnson and Mohapatra [45] for lower-triangular matrices A. They even allow certain more general spaces instead of l^p and l^q. However, their proofs are rather long (see [45, Theorems 6.7, 6.15, 6.16]). Indeed, our proof also works if l^p and l^q are replaced by any solid quasi-BK-spaces with quasi-norms satisfying $|\boldsymbol{x}| \leq |\boldsymbol{y}| \Longrightarrow \|\boldsymbol{x}\| \leq \|\boldsymbol{y}\|$. This provides a positive answer to a question raised by Johnson and Mohapatra [45, p. 114].

Theorem 9.3 also holds for arbitrary, not necessarily row-finite matrices A if $p > 1$ (more generally, if the Köthe dual $E^\times$ of the quasi-BK-space E replacing l^p has the AK-property), as a slight variant of the above proof shows. However, the result fails for general A if $p = 1$ as shown by the example $A = (1), \boldsymbol{b} = \boldsymbol{1}$

and $q = \infty$. The same example also shows that compactness of $AD_{\boldsymbol{b}} : l^p \to l^q$ need not coincide with compactness of $D_{\boldsymbol{b}} : l^p \to A(q)$.

We now apply Theorem 9.3 to characterise compactness of factorable matrices.

Theorem 9.4 *let* $0 < p, q \le \infty$, *and suppose that the factorable matrix* $M(\boldsymbol{a}, \boldsymbol{b})$ *maps* l^p *into* l^q.

(a) *If* $p > q$, *then* $M(\boldsymbol{a}, \boldsymbol{b}) : l^p \to l^q$ *is compact for all* $\boldsymbol{a}$ *and* $\boldsymbol{b}$.

(b) *If* $p \le q$, *then* $M(\boldsymbol{a}, \boldsymbol{b}) : l^p \to l^q$ *is compact if and only if, as* $n \to \infty$,

$$\begin{aligned}
\left(\sum_{k=n}^{\infty} a_k^q\right)^{1/q} \left(\sum_{k=1}^{n} b_k^{p^*}\right)^{1/p^*} &\to 0 \qquad \textit{if} \ \ 1 < p \le q < \infty, \\
b_n \left(\sum_{k=n}^{\infty} a_k^q\right)^{1/q} &\to 0 \qquad \textit{if} \ \ p \le 1, p \le q < \infty, \\
a_n \left(\sum_{k=1}^{n} b_k^{p^*}\right)^{1/p^*} &\to 0 \qquad \textit{if} \ \ 1 < p; q = \infty, \boldsymbol{a} \notin l^\infty \setminus c_0, \\
\sum_k b_k^{p^*} &< \infty \qquad \textit{if} \ \ 1 < p; q = \infty, \boldsymbol{a} \in l^\infty \setminus c_0, \\
b_n \sup_{k \ge n} a_k &\to 0 \qquad \textit{if} \ \ p \le 1, q = \infty.
\end{aligned}$$

Proof. We can assume that $\boldsymbol{a} \notin \varphi$. Applying Theorem 9.3 with $A = M(\boldsymbol{a}, \mathbf{1})$ we see that $M(\boldsymbol{a}, \boldsymbol{b}) = M(\boldsymbol{a}, \mathbf{1}) D_{\boldsymbol{b}}$ maps l^p compactly into l^q if and only if $P_n \boldsymbol{b} \to \boldsymbol{b}$ in $\mathcal{M}(l^p, c(\boldsymbol{a}, 1, q))$ with respect to the operator topology. By Observation 9.1 the latter multiplier space is described by Bennett's conditions (i) - (viii) of **BIII**, Theorem 1 and by (9.2ix) above, hence coincides with $c(\boldsymbol{g}, p \to 1, p \to q)$ for some $\boldsymbol{g}$ (see also Theorem 7.7). Since the multipliers form a closed subspace in the space of all operators under the operator norm, and since no sequence space carries more than one Fréchet K-space topology [99, 5-5-8], we deduce that $P_n \boldsymbol{b} \to \boldsymbol{b}$ in $\mathcal{M}(l^p, c(\boldsymbol{a}, 1, q))$ if and only if $\boldsymbol{b} \in c(\boldsymbol{g}, p \to 1, p \to q)_{AK}$. Now, (a) follows from the fact that $c(\boldsymbol{g}, u, v)$ has the AK-property for $v < \infty$, and (b) follows with Propositions 5.1 and 6.1. □

Part (a) will come as no surprise. In fact, every matrix operator from l^p into l^q is compact if $0 < q < p \le \infty$. Such a result was first obtained by Littlewood

[69] who proved the case $q = 1, p = \infty$. For $q \geq 1, p < \infty$ the result is due to Pitt [83], for $1 < q < \infty, p = \infty$ to Sargent [85, p. 85, (d)] and for $q < 1$ to Stiles [90]. Also, part (b) follows from general results in the cases where $p = 1$ or $q = \infty$ [85, p. 85, (a), (b)].

Theorem 9.4 in particular characterises the compactness of Rhaly matrices $R_{\boldsymbol{a}} = M(\boldsymbol{a}, \mathbf{1})$, thereby completing the investigation of Rhaly [84] and Leibowitz [57] mentioned above.

The results obtained in this section so far have analogues for the transposes of the factorable matrices $M(\boldsymbol{a}, \boldsymbol{b})$. With **BIII** we let $N(\boldsymbol{a}, \boldsymbol{b})$ denote the matrix with entries

$$a_{nk} = \begin{cases} a_n b_k & \text{if } k \geq n \\ 0 & \text{if } k < n, \end{cases}$$

where $\boldsymbol{a}$ and $\boldsymbol{b}$ are non-negative sequences, and $N(\boldsymbol{a}, \boldsymbol{b})$ will also be called an *(upper-triangular) factorable matrix.* The inequalities corresponding to the l^p-l^q-mapping problem for $N(\boldsymbol{a}, \boldsymbol{b})$ read

$$(9.4) \qquad \left(\sum_n \left(a_n \sum_{k=n}^{\infty} b_k x_k \right)^q \right)^{1/q} \leq K \left(\sum_n x_n^p \right)^{1/p}.$$

The earliest inequalities of this type were given by Copson [27] in 1927, a particular case of which is the inequality

$$(9.5) \qquad \sum_n \left(\sum_{k=n}^{\infty} \frac{x_k}{k} \right)^p \leq K \sum_n x_n^p$$

for $1 \leq p < \infty$, compare [40, Theorem 331]. Incidentally, the constant $K = p^p$ turns out to be best-possible.

Since $N(\boldsymbol{a}, \boldsymbol{b})$ is simply the transpose $M(\boldsymbol{b}, \boldsymbol{a})^t$, often a result on N follows from a corresponding result on M by abstract principles. Thus, for example, it was already noted by Hardy [38] that inequality (9.5) can be deduced from his inequality (0.1) and vice versa by what Bennett calls the *method of transposes* (**BI**, p. 410).

However, since such a reduction does not always seem to be possible, an independent study of the matrices $N(\boldsymbol{a}, \boldsymbol{b})$ alongside $M(\boldsymbol{a}, \boldsymbol{b})$ is called for. We first treat the l^p-l^q-mapping problem or, equivalently, the inequalities (9.4).

Observation 9.5 *Let* $0 < p, q \leq \infty$ *and* $\boldsymbol{a} \neq \mathbf{0}$. *Then a factorable matrix* $N(\boldsymbol{a}, \boldsymbol{b})$ *maps* l^p *into* l^q *if and only if* $\boldsymbol{b}$ *is a multiplier from* l^p *into* $d(\boldsymbol{a}, 1, q)$.

Here, and in the sequel, we need to exclude the case $\boldsymbol{a} = \mathbf{0}$, compare the remark after Proposition 5.1.

The characterising conditions now follow immediately from Theorem 7.13 and Remark 7.14. It turns out that they can be obtained from the conditions for $M(\boldsymbol{a}, \boldsymbol{b})$ by a simple procedure. Let us denote conditions (i) to (viii) in Theorem 1 of **BIII** by (9.2i) to (9.2viii); (9.2ix) and (9.2x) are given above. Now denote by (9.2i') to (9.2x') the same conditions after replacing any summation or supremum on $[1, \nu]$ by one on $[\nu, \infty)$, $\nu = N, n, k$, and vice versa.

Theorem 9.6 *Let $0 < p, q \leq \infty$ and $\boldsymbol{a} \neq \mathbf{0}$. Then a factorable matrix $N(\boldsymbol{a}, \boldsymbol{b})$ maps l^p into l^q if and only if conditions (9.2i') to (9.2x') hold (depending on the value of p and q).*

For example, if $0 < q < p \leq 1$, then $N(\boldsymbol{a}, \boldsymbol{b})$ maps l^p into l^q if and only if

(9.2ix') $$\sum_n a_n^q \left(\sum_{k=1}^n a_k^q \right)^{r/p} \sup_{k \geq n} b_k^r < \infty,$$

where $\frac{1}{r} = \frac{1}{q} - \frac{1}{p}$.

We next turn again to the question when $N(\boldsymbol{a}, \boldsymbol{b}) : l^p \to l^q$ is compact. Theorem 9.3 is not applicable here, at least for $p \leq 1$, since $N(\boldsymbol{a}, \boldsymbol{b})$ is not row-finite (compare the remarks following the theorem). The proof of the following characterisation is less satisfactory than that of Theorem 9.4.

Theorem 9.7 *Let $0 < p, q \leq \infty$, $\boldsymbol{a} \neq \mathbf{0}$, and suppose that the factorable matrix $N(\boldsymbol{a}, \boldsymbol{b})$ maps l^p into l^q.*

(a) *If $p > q$, then $N(\boldsymbol{a}, \boldsymbol{b}) : l^p \to l^q$ is compact for all $\boldsymbol{a}$ and $\boldsymbol{b}$.*

(b) *If $p \leq q$, then $N(\boldsymbol{a}, \boldsymbol{b}) : l^p \to l^q$ is compact if and only if, as $n \to \infty$,*

$$\left(\sum_{k=1}^n a_k^q \right)^{1/q} \left(\sum_{k=n}^\infty b_k^{p^*} \right)^{1/p^*} \to 0 \quad \text{if} \quad 1 < p \leq q < \infty,$$

$$b_n \left(\sum_{k=1}^n a_k^q \right)^{1/q} \to 0 \quad \text{if} \quad p \leq 1, p \leq q < \infty, \boldsymbol{b} \notin l^\infty \setminus c_0,$$

$$\sum_k a_k^q < \infty \quad \text{if} \quad p \leq 1, p \leq q < \infty, \boldsymbol{b} \in l^\infty \setminus c_0,$$

$$a_n \left(\sum_{k=n}^{\infty} b_k^{p^*}\right)^{1/p^*} \to 0 \qquad \textit{if} \quad 1 < p; q = \infty,$$

$$a_n \sup_{k \geq n} b_k \to 0 \qquad \textit{if} \quad p \leq 1, q = \infty.$$

Proof. (a) follows from the theorem of Pitt, Stiles and others mentioned after Theorem 9.4 above.

(b) If $p = q = \infty$, the result is a special case of a general theorem, compare [85, p. 85, (b)]. If $\boldsymbol{b} \in \varphi$, (b) is trivial. Hence we assume that $p < \infty$ or $q < \infty$, and fix $\boldsymbol{b} \notin \varphi$. The sequence space $X = \{\boldsymbol{a} : N(\boldsymbol{a}, \boldsymbol{b}) : l^p \to l^q\}$ can be quasi-normed by the operator norm, that is, $\|\boldsymbol{a}\|_X = \sup\{(\sum_n |a_n \sum_{k=n}^{\infty} b_k x_k|^q)^{1/q} : \|\boldsymbol{x}\|_{l^p} \leq 1\}$. We claim that $N(\boldsymbol{a}, \boldsymbol{b})$ is compact if and only if $\|\boldsymbol{a} - P_n \boldsymbol{a}\|_X \to 0$. If $q < \infty$, this follows from the well-known characterisation of relatively compact subsets of l^q. If $p < q = \infty$, we can argue that, since $N(\boldsymbol{a}, \boldsymbol{b})$ is column-finite, it maps φ into c_0, hence also l^p into c_0, and the claim follows from the characterisation of relatively compact subsets of c_0.

It follows from Theorem 9.6 that, for $p \leq q$, X is of the form $c(\boldsymbol{h}, q, \infty)$ for some $\boldsymbol{h}$ depending on $\boldsymbol{b}$. As in the proof of Theorem 9.4 we deduce that $\|\boldsymbol{a} - P_n \boldsymbol{a}\|_X \to 0$ if and only if $\boldsymbol{a} \in c(\boldsymbol{h}, q, \infty)_{AK}$. Thus (b) follows with Propositions 5.1 and 6.1. □

Clearly we could have studied factorable matrices $M(\boldsymbol{a}, \boldsymbol{b})$ and $N(\boldsymbol{a}, \boldsymbol{b})$ for general real or complex sequences $\boldsymbol{a}$ and $\boldsymbol{b}$. However, if is not difficult to see that a general matrix $M(\boldsymbol{a}, \boldsymbol{b})$ maps l^p into l^q if and only if $M(|\boldsymbol{a}|, |\boldsymbol{b}|)$ does, and $M(\boldsymbol{a}, \boldsymbol{b}) : l^p \to l^q$ is compact if and only if $M(|\boldsymbol{a}|, |\boldsymbol{b}|)$ is (see the proof of Theorem 9.3). The same is true for $N(\boldsymbol{a}, \boldsymbol{b})$.

Closing this section we remark that we have not made full use of Theorems 7.7, 7.13 and their remarks. Theorem 7.7 and Remark 7.8 yield a characterisation of the sequences $\boldsymbol{a}$ and $\boldsymbol{b}$ for which the inequality

$$\left(\sum_n \left[a_n \left(\sum_{k=1}^{n} (b_k x_k)^p\right)^{1/p}\right]^q\right)^{1/q} \leq K \left(\sum_n x_n^r\right)^{1/r}$$

holds with some $K > 0$, for all non-negative sequences $\boldsymbol{x}$; Theorem 7.13 and Remark 7.14 do the same for the inequality

$$\left(\sum_n \left[a_n \left(\sum_{k=n}^{\infty} (b_k x_k)^p\right)^{1/p}\right]^q\right)^{1/q} \leq K \left(\sum_n x_n^r\right)^{1/r}.$$

10 Lower Bounds for Factorable Matrices and Copson's Inequality

In the previous section we have studied weighted generalisations of Hardy's inequality

$$\sum_n \left(\frac{1}{n}\sum_{k=1}^{n} x_k\right)^p \le K \sum_n x_n^p$$

and its companion

$$\sum_n \left(\sum_{k=n}^{\infty} \frac{x_k}{k}\right)^p \le K \sum_n x_n^p,$$

see (9.5). These results are only valid for $p > 1$ and $p \ge 1$, respectively. In many cases such an inequality remains true for $p < 1$ if the sign of inequality is reversed, compare [40, p. 250]. This is trivially so for Hardy's inequality because then the left-hand side diverges for $\boldsymbol{x} \ne \mathbf{0}$. On the other hand, although the second inequality is equivalent to Hardy's (see Section 9), it turns into a new and non-trivial inequality for $p < 1$:

$$\sum_n \left(\sum_{k=n}^{\infty} \frac{x_k}{k}\right)^p \ge K \sum_n x_n^p \tag{10.1}$$

holds with some $K > 0$ for all non-negative sequences $\boldsymbol{x}$. This is known as ***Copson's inequality*** [28], compare [40, Theorem 344]. Here, the constant $K = p^p$ is best-possible.

In **BII**, Bennett uses the following terminology to deal with inequalities like Copson's. A matrix A with non-negative entries is called $p \to q$ ***bounded below*** if there is some constant $K > 0$ such that

$$||A\boldsymbol{x}||_{l^q} \ge K||\boldsymbol{x}||_{l^p}$$

for all non-negative sequences $\boldsymbol{x}$. We do not exclude the trivial case where $||A\boldsymbol{x}||_{l^q} = \infty$ for all $\boldsymbol{x} \ne \mathbf{0}$ as, for instance, in Hardy's reversed inequality. The largest constant K, denoted by $L(A)$, is called the $p \to q$ ***lower bound*** of A. In this language, Copson's inequality can be viewed as asserting the $p \to p$ lower boundedness of the factorable matrix $N(\mathbf{1}, \frac{1}{k})$ for $p < 1$, and the lower bound turns out to be p.

A problem parallel to the matrix mapping problem mentioned at the beginning of the previous section is that of characterising lower boundedness of general matrices A in terms of their entries. This problem was solved by Bennett if $0 < q \leq p$ and $p \geq 1$ [11, Theorem 3], but it seems to be open and comparable in difficulty to the matrix mapping problem if $0 < p, q < 1$. Thus one turns again to subclasses of matrices. In **BII**, Bennett completely solved the lower boundedness problem for factorable matrices $M(\boldsymbol{a}, \boldsymbol{b})$ and $N(\boldsymbol{a}, \boldsymbol{b})$ if $0 < q \leq p < 1$ (**BII**, Theorem 4), thereby generalising Copson's inequality. He also gives some estimates on the lower bounds of these matrices. But he mentions that the case $0 < p < q < 1$ remains open (**BII**, p. 393).

Now, our work in Chapter II enables us to give a complete solution of the lower boundedness problem for factorable matrices over the whole range of p and q. This solves Bennett's problem and also gives a new and elementary proof of his characterisation in **BII**. On the other hand, once again the corresponding quantitative problem of finding useful estimates for lower bounds is outside the reach of our method.

We first consider the lower-triangular factorable matrices $M(\boldsymbol{a}, \boldsymbol{b})$ with non-negative sequences $\boldsymbol{a}$ and $\boldsymbol{b}$. The $p \to q$ lower boundedness problem consists in characterising when there is a constant $K > 0$ such that

$$\left(\sum_n \left(a_n \sum_{k=1}^n b_k x_k\right)^q\right)^{1/q} \geq K \left(\sum_n x_n^p\right)^{1/p} \tag{10.2}$$

for all non-negative sequences $\boldsymbol{x}$. This inequality fails whenever some $b_k = 0$. Thus, the sequence $\boldsymbol{b}$ has to be positive, so that $\boldsymbol{b}^{-1} = (1/b_k)_k$ exists. Now the closed graph theorem and a glance at Proposition 5.1 implies the following.

Observation 10.1 *Let $0 < p, q \leq \infty$. Then a factorable matrix $M(\boldsymbol{a}, \boldsymbol{b})$ is $p \to q$ bounded below if and only if $\boldsymbol{b}$ is positive and $\boldsymbol{b}^{-1}$ is a multiplier from $c(\boldsymbol{a}, 1, q)$ into l^p.*

Clearly the inequality (10.2) holds trivially for positive $\boldsymbol{b}$ if $\boldsymbol{a} \notin l^q$. For $\boldsymbol{a} \in l^q$, Theorem 7.2 and Remark 7.3 provide us with a complete list of characterisations. We refer to Remark 2.2(i) for the proper interpretation of these conditions.

Theorem 10.2 *Let $0 < p, q \le \infty$ and $\boldsymbol{a} \in l^q$. Then a factorable matrix $M(\boldsymbol{a}, \boldsymbol{b})$ is $p \to q$ bounded below if and only if $\boldsymbol{a} \notin \varphi$, $\boldsymbol{b}$ is positive and*

(10.3i)(a) $\inf_n \left(\sum_{k=n}^{\infty} a_k^q\right)^{1/q} \left(\sum_{k=n}^{\infty} b_k^{p^*}\right)^{1/p^*} > 0$ *or*

(10.3i)(b) $\sum_{k=n}^{\infty} \left(b_k \sum_{j=k}^{\infty} a_j^q\right)^{p^*} = O\left(\sum_{k=n}^{\infty} a_k^q\right)^{p^*/q^*}$ *if* $q \le p < 1$,

(10.3ii) $\inf_n b_n \left(\sum_{k=n}^{\infty} a_k^q\right)^{1/q} > 0$ *if* $q \le p; 1 \le p; q < \infty$,

(10.3iii) $\inf_n b_n \sup_{k \ge n} a_k > 0$ *if* $p = q = \infty$,

(10.3iv)(a) $\sum_n \left(\sum_{k=n}^{\infty} a_k^q\right)^{-r/q} b_n^{p^*} \left(\sum_{k=n}^{\infty} b_k^{p^*}\right)^{-r/q^*} < \infty$ *or*

(10.3iv)(b) $\sum_n a_n^q \left(\sum_{k=1}^{n} \left[b_k \left(\sum_{j=k}^{\infty} a_j^q\right)^{1/p}\right]^{p^*}\right)^{-r/p^*} < \infty$ *or*

(10.3iv)(c) $\sum_n a_{n-1}^q \left(\sum_{k=n-1}^{\infty} a_k^q\right)^{-1} \left(\sum_{k=n}^{\infty} a_k^q\right)^{-r/q} \left(\sum_{k=n}^{\infty} b_k^{p^*}\right)^{-r/p^*} < \infty$

if $p < q < \infty, p < 1$,

(10.3v)(a) $\sum_n a_n^q \sup_{k \le n} \left(b_k \left(\sum_{j=k}^{\infty} a_j^q\right)^{1/p}\right)^{-r} < \infty$ *or*

(10.3v)(b) $\sum_n a_{n-1}^q \left(\sum_{k=n-1}^{\infty} a_k^q\right)^{-1} \left(\sum_{k=n}^{\infty} a_k^q\right)^{-r/q} \sup_{k \ge n} b_k^{-r} < \infty$

if $1 \le p < q < \infty$,

(10.3vi) $\sum_n \left(\sup_{k \ge n} a_k\right)^{-p} b_n^{p*} \left(\sum_{k=n}^{\infty} b_k^{p^*}\right)^{-p} < \infty$ *if* $p < 1, q = \infty$,

(10.3vii) $\sum_n \left(\left(\sup_{k \ge n} a_k\right)^{-p} - \left(\sup_{k \ge n-1} a_k\right)^{-p}\right) \sup_{k \ge n} b_k^{-p} < \infty$

if $1 \le p < q = \infty$,

where $\frac{1}{r} = \frac{1}{p} - \frac{1}{q}$.

The conditions (10.3iv) solve the problem posed by Bennett in **BII**, p. 393.

As mentioned above, the conditions (10.3i) are due to Bennett, while (10.3ii) is a special case of a general result, cf. [11, Theorem 3]. For the case $q = 1$ in (10.3iv) see also Proposition 10.6 below.

We turn to the lower boundedness problem for the upper-triangular factorable matrices $N(\boldsymbol{a}, \boldsymbol{b})$ with non-negative sequences $\boldsymbol{a}$ and $\boldsymbol{b}$, that is, the inequality

$$\left(\sum_n \left(a_n \sum_{k=n}^{\infty} b_k x_k\right)^q\right)^{1/q} \geq K \left(\sum_n x_n^p\right)^{1/p} \tag{10.4}$$

with some constant $K > 0$ for all non-negative sequences $\boldsymbol{x}$. We can regard this as a *weighted Copson inequality.*

Clearly, (10.4) can only hold if a_1 and $\boldsymbol{b}$ are positive. Hence the closed graph theorem and Proposition 5.1 imply the following.

Observation 10.3 *Let* $0 < p, q \leq \infty$. *Then a factorable matrix* $N(\boldsymbol{a}, \boldsymbol{b})$ *is* $p \to q$ *bounded below if and only if* a_1 *and* $\boldsymbol{b}$ *are positive and* $\boldsymbol{b}^{-1}$ *is a multiplier from* $d(\boldsymbol{a}, 1, q)$ *into* l^p.

In view of Proposition 5.1 we can concentrate on $\boldsymbol{a} \notin l^q$. In that case, the characterising conditions follow from Theorem 7.9 and Remark 7.10. They can be obtained from the conditions (10.3i) to (10.3vii) by simply replacing any summation or supremum on $[\nu, \infty)$ by one on $[1, \nu], \nu = n, k$, and vice versa, by replacing $[n-1, \infty)$ by $[1, n+1]$ and a_{n-1} by a_{n+1}. We call the new conditions (10.3i') to (10.3vii').

Theorem 10.4 *Let* $0 < p, q \leq \infty$ *and* $\boldsymbol{a} \notin l^q$. *Then a factorable matrix* $N(\boldsymbol{a}, \boldsymbol{b})$ *is* $p \to q$ *bounded below if and only if* a_1 *and* $\boldsymbol{b}$ *are positive and conditions* (10.3i') *to* (10.3vii') *hold (depending on* p *and* q*).*

Thus, for example, for $p < q < \infty$ and $p < 1$, $N(\boldsymbol{a}, \boldsymbol{b})$ is $p \to q$ bounded below if and only if $a_1 > 0, \boldsymbol{b}$ is positive and

$$\sum_n a_n^q \left(\sum_{k=n}^{\infty} \left[b_k \left(\sum_{j=1}^{k} a_j^q\right)^{1/p}\right]^{p^*}\right)^{-r/p^*} < \infty,$$

where $\frac{1}{r} = \frac{1}{p} - \frac{1}{q}$. We remark that one may allow $\boldsymbol{a} \in l^q$ here.

In their study of power series with positive coefficients and of moment constants for positive functions Hardy and Littlewood [39, Theorem 1] have established four inequalities that contain Hardy's and Copson's inequality as special cases. Let c be a real parameter and $0 < p < \infty$. Then

$$\text{(10.5i)} \quad \sum_n n^{-c} \left(\sum_{k=1}^{n} x_k \right)^p \leq K \sum_n n^{-c} (n x_n)^p \qquad \text{for} \quad p, c > 1,$$

$$\text{(10.5ii)} \quad \sum_n n^{-c} \left(\sum_{k=n}^{\infty} x_k \right)^p \leq K \sum_n n^{-c} (n x_n)^p \qquad \text{for} \quad c < 1 < p,$$

$$\text{(10.5iii)} \quad \sum_n n^{-c} \left(\sum_{k=1}^{n} x_k \right)^p \geq K \sum_n n^{-c} (n x_n)^p \qquad \text{for} \quad p < 1 < c,$$

$$\text{(10.5iv)} \quad \sum_n n^{-c} \left(\sum_{k=n}^{\infty} x_k \right)^p \geq K \sum_n n^{-c} (n x_n)^p \qquad \text{for} \quad p, c < 1$$

with a positive constant K depending on p and c. It is remarkable that the best-possible constants are only known for certain values of p and c, see **BI**, **BII**, [28] and [66, D. 62]. The inequalities of Hardy and Littlewood involve the special factorable matrices $M(\boldsymbol{a}, \boldsymbol{b})$ and $N(\boldsymbol{a}, \boldsymbol{b})$ with

$$a_n = n^{-c/p}, \; b_k = k^{c/p-1}.$$

This motivated Bennett to study the l^p-l^q-mapping problem and the $p \longrightarrow q$ lower boundedness problem for the class of factorable matrices $M(\boldsymbol{a}, \boldsymbol{b})$ and $N(\boldsymbol{a}, \boldsymbol{b})$ with

$$a_n = n^{-\alpha}, \; b_k = k^{-\beta}$$

for $\alpha, \beta \in \mathbb{R}$. The solution of the first problem is given (or hinted at) in **BIII**, p. 161. The second problem is only solved for $0 < p, q \leq 1$ in **BII**, Theorem 5. The following result, a corollary of Theorems 10.2 and 10.4, completes Bennett's investigation. As usual we set $1/\infty = 0$.

Theorem 10.5 *Let* $0 < p, q \leq \infty$ *and* $\alpha, \beta \in \mathbb{R}$.
(a) *The factorable matrix* $M(n^{-\alpha}, k^{-\beta})$ *is* $p \to q$ *bounded below if and only if*

$$
\begin{array}{lll}
\alpha \leq \frac{1}{q} & \text{or} \quad \alpha + \beta \leq \frac{1}{q} - \frac{1}{1 \to p} & \text{if} \quad q \leq p; q < \infty, \\
\alpha < \frac{1}{q} & \text{or} \quad \alpha + \beta \leq \frac{1}{q} - \frac{1}{1 \to p} & \text{if} \quad p = q = \infty, \\
\alpha \leq \frac{1}{q} & \text{or} \quad \alpha + \beta < \frac{1}{q} - \frac{1}{1 \to p} & \text{if} \quad p < q < \infty, \\
\alpha < \frac{1}{q} & \text{or} \quad \alpha + \beta < \frac{1}{q} - \frac{1}{1 \to p} & \text{if} \quad p < 1; q = \infty, \\
\alpha < \frac{1}{q} & \text{or} \quad \alpha + \beta < \frac{1}{q} - \frac{1}{1 \to p} & \\
 & \text{or} \quad \alpha = \beta = 0 & \text{if} \quad 1 \leq p; q = \infty.
\end{array}
$$

(b) *The factorable matrix* $N(n^{-\alpha}, k^{-\beta})$ *is* $p \to q$ *bounded below if and only if*

$$
\begin{array}{lll}
\beta < -\frac{1}{1 \to p} & \text{or} \quad \alpha + \beta \leq \frac{1}{q} - \frac{1}{1 \to p} & \text{if} \quad q \leq p < 1, \\
\beta \leq -\frac{1}{1 \to p} & \text{or} \quad \alpha + \beta \leq \frac{1}{q} - \frac{1}{1 \to p} & \text{if} \quad q \leq p; 1 \leq p, \\
\beta < -\frac{1}{1 \to p} & \text{or} \quad \alpha + \beta < \frac{1}{q} - \frac{1}{1 \to p} & \text{if} \quad p \leq \frac{q}{q+1} < q < \infty \\
 & & \text{or} \quad p < 1; q = \infty, \\
\beta < -\frac{1}{1 \to p} & \text{or} \quad \alpha + \beta < \frac{1}{q} - \frac{1}{1 \to p} & \\
 & \text{or} \quad \alpha = \frac{1}{q}, \beta = -\frac{1}{1 \to p} & \text{if} \quad \frac{q}{q+1} < p < 1; p < q < \infty, \\
\beta \leq -\frac{1}{1 \to p} & \text{or} \quad \alpha + \beta < \frac{1}{q} - \frac{1}{1 \to p} & \text{if} \quad 1 \leq p < q \leq \infty.
\end{array}
$$

We wish to add a remark on a recent paper of Leindler [62] who also treats mapping properties of general factorable matrices complementing certain results of Bennett. This is seen after suitable changes of variables.

If we substitute $q = r, p = r/s, a_n = u_n^{1/r}, b_n = v_n^{(r-s)/r}$ (we may assume that the v_n are positive), $x_n = v_n^{s/r} w_n$ and let $N \to \infty$, we see that Leindler's Theorems 1 and 1' provide sufficient conditions for factorable matrices $N(\boldsymbol{a}, \boldsymbol{b})$ and $M(\boldsymbol{a}, \boldsymbol{b})$ to map l^p into l^q with norm ≤ 1, when $0 < q < p \leq 1$. The interest in this result lies in the fact that it provides a norm estimate. As for its qualitative aspect, the appropriate necessary and sufficient condition is given in

Theorems 9.2 and 9.6. Leindler also treats the case $0 < q = p \leq 1$, but this is already covered by a general result on arbitrary matrices (see (25) of **BIII**).

By the substitution $p = 1/s, a_n = u_n, b_n = v_n^{r-1/p}$ and $x_n = v_n^{1/p} w_n^r$ (we may assume $v_n > 0$ again; the parameter r may be kept) Leindler's Theorems 2 and 2' give sufficient conditions for the factorable matrices $M(\boldsymbol{a}, \boldsymbol{b})$ and $N(\boldsymbol{a}, \boldsymbol{b})$ to be $p \to 1$ bounded below with lower bound at least 1, when $0 < p \leq 1$. However, this problem can be completely solved for arbitrary matrices as follows. (We omit the trivial case $p = 1$.)

Proposition 10.6 *Let $0 < p < 1$. Then a (finite or infinite) non-negative matrix $A = (a_{nk})$ is $p \to 1$ bounded below if and only if*

$$\Delta := \sum_k \left(\sum_n a_{nk} \right)^{p^*} < \infty,$$

where we interpret $0^{p^*} = \infty$ and $\infty^{p^*} = 0$. In that case, $L(A) = \Delta^{1/p^*}$.

Proof. We have to characterise when there is some $K > 0$ such that

$$\|A\boldsymbol{x}\|_{l^1} \geq K \|\boldsymbol{x}\|_{l^p}$$

holds for all non-negative sequences $\boldsymbol{x}$. Changing the order of summation on the left shows that this is equivalent to $\left((\sum_n a_{nk})^{-1} \right)_k$ being a multiplier from l^1 to l^p of norm at most K^{-1}. The result then follows. □

Thus, in Leindler's Theorems 2 and 2' the necessary and sufficient conditions in the case $s > 1$ turn out to be, respectively,

$$\sum_{k=1}^N \left(v_k^{r-s} \sum_{n=k}^N u_n \right)^{1/(1-s)} \leq 1 \quad \text{and} \quad \sum_{k=1}^N \left(v_k^{r-s} \sum_{n=1}^k u_n \right)^{1/(1-s)} \leq 1.$$

11 Reverse Hardy-Copson-Higaki Inequalities

Hardy remarks that his inequality "naturally suggest[s] numerous generalisations". He proceeds to give the following as an example. Let $\boldsymbol{\lambda} = (\lambda_n)$ be a sequence of positive numbers and set $\Lambda_n = \sum_{k=1}^{n} \lambda_k$. Then, for $p > 1$, the inequality

$$\sum_n \lambda_n \Lambda_n^{-p} \left(\sum_{k=1}^{n} \lambda_k x_k \right)^p \leq K \sum_n \lambda_n x_n^p$$

holds with some $K > 0$ for all non-negative sequences $\boldsymbol{x}$ [36, Theorem C]. This inequality and the Hardy-Littlewood inequalities (10.5) stimulated further research which has led to the following group of four inequalities where $p > 1$ and $c \in \mathbb{R}$:

$$\text{(11.1i)} \qquad \sum_n \lambda_n \Lambda_n^{-c} \left(\sum_{k=1}^{n} \lambda_k x_k \right)^p \leq K \sum_n \lambda_n \Lambda_n^{p-c} x_n^p \qquad \text{for} \quad c > 1,$$

$$\text{(11.1ii)} \qquad \sum_n \lambda_n \Lambda_n^{-c} \left(\sum_{k=n}^{\infty} \lambda_k x_k \right)^p \leq K \sum_n \lambda_n \Lambda_n^{p-c} x_n^p \qquad \text{for} \quad c < 1,$$

$$\text{(11.1iii)} \qquad \sum_n \lambda_n (\Lambda_n^*)^{-c} \left(\sum_{k=1}^{n} \lambda_k x_k \right)^p \leq K \sum_n \lambda_n (\Lambda_n^*)^{p-c} x_n^p \qquad \text{for} \quad c < 1,$$

$$\text{(11.1iv)} \qquad \sum_n \lambda_n (\Lambda_n^*)^{-c} \left(\sum_{k=n}^{\infty} \lambda_k x_k \right)^p \leq K \sum_n \lambda_n (\Lambda_n^*)^{p-c} x_n^p \qquad \text{for} \quad c > 1,$$

where Λ_n^* is defined as $\sum_{k=n}^{\infty} \lambda_k$. In the sequel, whenever Λ_n^* appears it is assumed that $\sum_k \lambda_k < \infty$. For a derivation of these inequalities and a discussion of best-possible constants we refer the reader to Section 5 of **BI**. Bennett also observes that by virtue of the method of transposes and sinister transposes (cf. **BI**, pp. 408 and 410) the four inequalities are in fact emanations of a single result. (In contrast to what is said after Corollary 6 of **BI**, however, it seems that (11.1iii) is a sinister transpose of (11.1i) rather than of (11.1ii), and (11.1iv) is a sinister transpose of (11.1ii).)

As we have mentioned above, inequality (11.1i) is due to Hardy for $c = p$, and it is due to Copson [28] for $1 < c \leq p$, who also proved (11.1ii) for $0 \leq c < 1$, see in addition [37] and [27]. Inequalities (11.1iii) and (11.1iv) are due to Higaki [42] for $c = 0$ and $c = p$, respectively. Higaki's contribution seems to have been overlooked so far. In [58] Leindler independently obtained (11.1ii) and (11.1iii) for $c = 0$. All the remaining cases were first obtained by Bennett in **BI**. We shall refer to (11.1) as the *Hardy-Copson-Higaki inequalities.*

Now, in [61] Leindler has investigated the question of how much is lost in passing from the left to the right in the inequalities (11.1.ii) and (11.1iii) for $c = 0$, more precisely, for which $\boldsymbol{\lambda}$ these inequalities can be reversed. We extend here Leindler's result to cover (11.1i) - (11.1iv) for all values of c.

Theorem 11.1 *Let $p > 1$, and let $\boldsymbol{\lambda}$ be a positive sequence.*

(a) *For $c > 1$, the inequality*

$$\sum_n \lambda_n \Lambda_n^{-c} \left(\sum_{k=1}^{n} \lambda_k x_k \right)^p \geq K \sum_n \lambda_n \Lambda_n^{p-c} x_n^p$$

holds with some $K > 0$ if and only if (Λ_n/λ_n) is bounded.

(b) *For $c < 1$, the inequality*

$$\sum_n \lambda_n \Lambda_n^{-c} \left(\sum_{k=n}^{\infty} \lambda_k x_k \right)^p \geq K \sum_n \lambda_n \Lambda_n^{p-c} x_n^p$$

holds with some $K > 0$ if and only if (Λ_n/λ_n) is bounded.

(c) *For $c < 1$, the inequality*

$$\sum_n \lambda_n (\Lambda_n^*)^{-c} \left(\sum_{k=1}^{n} \lambda_k x_k \right)^p \geq K \sum_n \lambda_n (\Lambda_n^*)^{p-c} x_n^p$$

holds with some $K > 0$ if and only if (Λ_n^*/λ_n) is bounded.

(d) *For $c > 1$, the inequality*

$$\sum_n \lambda_n (\Lambda_n^*)^{-c} \left(\sum_{k=n}^{\infty} \lambda_k x_k \right)^p \geq K \sum_n \lambda_n (\Lambda_n^*)^{p-c} x_n^p$$

holds with some $K > 0$ if and only if (Λ_n^*/λ_n) is bounded.

Proof. We show (c). The inequality can be viewed as asserting that the factorable matrix $M(\lambda_n^{1/p}(\Lambda_n^*)^{-c/p}, \lambda_k^{1-1/p}(\Lambda_k^*)^{c/p-1})$ is $p \to p$ bounded below. By Theorem 10.2 this is equivalent to

$$\lambda_n^{1-p}(\Lambda_n^*)^{p-c} \leq K \sum_{k=n}^{\infty} \lambda_k (\Lambda_k^*)^{-c} \qquad (n \in \mathbb{N})$$

for some $K > 0$. Since $c < 1$, this is equivalent to

$$\lambda_n^{1-p}(\Lambda_n^*)^{p-c} \leq K(\Lambda_n^*)^{-c+1}$$

by (2.6ii), hence to the boundedness of (Λ_n^*/λ_n). The proof of (b) is similar.

The inequality given in (a) is equivalent to

$$\lambda_n^{1-p} \Lambda_n^{p-c} \leq K \sum_{k=n}^{\infty} \lambda_k \Lambda_k^{-c} \qquad (n \in \mathbb{N}) \tag{11.2}$$

for some $K > 0$. Since $c > 1$, we have

$$\sum_{k=n}^{\infty} \lambda_k \Lambda_k^{-c} \leq \int_{\Lambda_{n-1}}^{\infty} t^{-c} dt = K \Lambda_{n-1}^{-c+1},$$

so that (11.2) together with $\Lambda_{n-1} = \Lambda_n - \lambda_n$ implies that

$$\left(\frac{\Lambda_n}{\lambda_n}\right)^{p-1} \left(1 - \frac{\lambda_n}{\Lambda_n}\right)^{c-1} \leq K.$$

This shows that (Λ_n/λ_n) has to be bounded. Conversely, if the latter it true, then

$$\lambda_n^{1-p} \Lambda_n^{p-c} = \left(\frac{\Lambda_n}{\lambda_n}\right)^p \lambda_n \Lambda_n^{-c} \leq K \sum_{k=n}^{\infty} \lambda_k \Lambda_k^{-c} \qquad (n \in \mathbb{N})$$

gives (11.2). The proof of (d) is similar. □

It is interesting to note that the characterising condition on $\boldsymbol{\lambda}$ is independent of the parameters c and p.

In order to see that our result contains Leindler's as a special case one needs to take account of his Lemma in [61]. Although new, Theorem 11.1 cannot strictly be called an application of our work in Chapters I and II. Its proof can be given just as well by using Bennett's characterisation of $p \to p$ lower boundedness of general matrices for $p \geq 1$ given in [11, Theorem 3] or by suitably modifying Leindler's proof. In addition, Bennett's result even gives the best-possible constants K. For instance, in (a) it is

$$K = \inf_n \lambda_n^{1-1/p} \Lambda_n^{c/p-1} \left(\sum_{k=n}^{\infty} \lambda_k \Lambda_k^{-c}\right)^{1/p}.$$

12 Reverse Copson-Takahashi Inequalities, and a New Inequality

In this section we want to study analogues of the Hardy-Copson-Higaki inequalities for $p < 1$. As in Section 10 one might expect to obtain valid inequalities if the sign of inequality is reversed. This is indeed the case, at least under certain restrictions. We have the following group of four inequalities for $0 < p < 1$ and $c \in \mathbb{R}$:

If $\Lambda_n \to \infty$ and $\lambda_{n+1} \leq M\lambda_n$ for all n, then

$$\text{(12.1i)} \qquad \sum_n \lambda_n \Lambda_n^{-c} \left(\sum_{k=1}^{n} \lambda_k x_k \right)^p \geq K \sum_n \lambda_n \Lambda_n^{p-c} x_n^p \qquad \text{for} \quad c > 1;$$

for all $\boldsymbol{\lambda}$,

$$\text{(12.1ii)} \qquad \sum_n \lambda_n \Lambda_n^{-c} \left(\sum_{k=n}^{\infty} \lambda_k x_k \right)^p \geq K \sum_n \lambda_n \Lambda_n^{p-c} x_n^p \qquad \text{for} \quad c < 1;$$

for all $\boldsymbol{\lambda}$,

$$\text{(12.1iii)} \quad \sum_n \lambda_n (\Lambda_n^*)^{-c} \left(\sum_{k=1}^{n} \lambda_k x_k \right)^p \geq K \sum_n \lambda_n (\Lambda_n^*)^{p-c} x_n^p \qquad \text{for} \quad c < 1;$$

if $\lambda_n \leq M\lambda_{n+1}$ for all n, then

$$\text{(12.1iv)} \quad \sum_n \lambda_n (\Lambda_n^*)^{-c} \left(\sum_{k=n}^{\infty} \lambda_k x_k \right)^p \geq K \sum_n \lambda_n (\Lambda_n^*)^{p-c} x_n^p \qquad \text{for} \quad c > 1.$$

Throughout, M is a positive constant.

The analogy with the inequalities (11.1) has its limits as is seen, for instance, by the need to introduce restrictions on $\boldsymbol{\lambda}$. Also, these inequalities, unlike (11.1), do not seem to be equivalent to each other in some sense.

Inequalities (12.1i) and (12.1ii) are due to Copson [28]. For a good discussion of these inequalities and best-possible constants we refer to Section 3 of **BII** (the constant in Corollary 3 should read $pL/(c-1)$). Inequality (12.1iii) with $c = 0$ can be found in another neglected paper, [93] of Takahashi. For $c = 0$, (12.1ii) and (12.1iii) were also obtained independently by Leindler [58], [60]. We shall refer to (12.1) as the *Copson-Takahashi inequalities.*

Inequality (12.1iv) seems to be new (even for $c = p$), as is Takahashi's inequality (12.1iii) for values of c other than 0. We shall derive these inequalities here. Departing from the general practice in these notes we shall even give an estimate for the constants K involved. For this we have to rely on results of Bennett in **BII**.

Theorem 12.1 *Let $0 < p < 1$, and let $\boldsymbol{\lambda}$ be a positive sequence.*
(a) *If $c < 1$, then*

$$\sum_n \lambda_n (\Lambda_n^*)^{-c} \left(\sum_{k=1}^{n} \lambda_k x_k \right)^p \geq K \sum_n \lambda_n (\Lambda_n^*)^{p-c} x_n^p$$

with $K = (\frac{p}{1-c})^p$ for $c \leq 0$ and $K = p^p$ for $0 \leq c < 1$. The constant for $c \leq 0$ is best-possible.
(b) *Let $c > 1$. If $\lambda_n \leq M \lambda_{n+1}$ $(n \in \mathbb{N})$ for some $M > 0$, then*

$$\sum_n \lambda_n (\Lambda_n^*)^{-c} \left(\sum_{k=n}^{\infty} \lambda_k x_k \right)^p \geq \left(\frac{pL}{M(c-1)} \right)^p \sum_n \lambda_n (\Lambda_n^*)^{p-c} x_n^p,$$

where $L = 1 - (1 + M\lambda_1 / \Lambda_1^*)^{1-c}$.

Proof. (a) It suffices to apply Corollary 1 of **BII** to N-tuples $\boldsymbol{\lambda}'$ and $\boldsymbol{x}'$ obtained form $\boldsymbol{\lambda}$ and $\boldsymbol{x}$ by reversing the order of the coordinates, and then to let $N \to \infty$. To see that $(\frac{p}{1-c})^p$ ist best-possible for $c \leq 0$ if suffices, as usual, to consider $\lambda_n = 1/n^\alpha$ and $x_k = 1/k^\beta$ for suitable α and β. We omit the details.

(b) We apply Theorem 1 of **BII** with $r = p, s = 1, u_n = \lambda_n (\Lambda_n^*)^{-c}, v_n = R^{p/(1-p)} \lambda_n (\Lambda_n^*)^{(p-c)/(1-p)}$ and $w_n = R^{-p/(1-p)} (\Lambda_n^*)^{(c-p)/(1-p)} x_n$, where $R = \frac{L}{M(c-1)}$. Under these substitutions, (3) of **BII** gives the desired inequality after letting $N \to \infty$. Thus we need only check (2) of **BII**. To this end we set $\lambda_0 = M\lambda_1$ and $\Lambda_0^* = \lambda_0 + \Lambda_1^*$. Then we obtain for $n \geq 1$

$$\sum_{k=1}^{n} u_k = \sum_{k=1}^{n} \lambda_k (\Lambda_k^*)^{-c} \geq M^{-1} \sum_{k=1}^{n} \lambda_{k-1} (\Lambda_k^*)^{-c} \geq M^{-1} \int_{\Lambda_n^*}^{\Lambda_0^*} t^{-c} dt$$

$$\geq \frac{M^{-1}}{c-1} \left(1 - \left(\frac{\Lambda_0^*}{\Lambda_1^*} \right)^{1-c} \right) (\Lambda_n^*)^{1-c} = R (\Lambda_n^*)^{1-c},$$

which immediately implies (2). □

We now ask for which $\boldsymbol{\lambda}$ the Copson-Takahashi inequalities hold with reversed sign of inequality. Such a characterisation has already been obtained by Leindler [61] for the inequalities (12.1ii) and (12.1iii) with $c = 0$. His result is contained in the following.

Theorem 12.2 *Let $0 < p < 1$, and let $\boldsymbol{\lambda}$ be a positive sequence.*

(a) *If $\Lambda_n \to \infty$ and $\lambda_{n+1} \le M\lambda_n$ for all n, then, for $c > 1$, the inequality*

$$\sum_n \lambda_n \Lambda_n^{-c} \left(\sum_{k=1}^{n} \lambda_k x_k \right)^p \le K \sum_n \lambda_n \Lambda_n^{p-c} x_n^p$$

holds with some $K > 0$ if and only if (Λ_n/λ_n) is bounded.

(b) *For $c < 1$, the inequality*

$$\sum_n \lambda_n \Lambda_n^{-c} \left(\sum_{k=n}^{\infty} \lambda_k x_k \right)^p \le K \sum_n \lambda_n \Lambda_n^{p-c} x_n^p$$

holds with some $K > 0$ if and only if (Λ_n/λ_n) is bounded.

(c) *For $c < 1$, the inequality*

$$\sum_n \lambda_n (\Lambda_n^*)^{-c} \left(\sum_{k=1}^{n} \lambda_k x_k \right)^p \le K \sum_n \lambda_n (\Lambda_n^*)^{p-c} x_n^p$$

holds with some $K > 0$ if and only if (Λ_n^*/λ_n) is bounded.

(d) *If $\lambda_n \le M\lambda_{n+1}$ for all n, then, for $c > 1$, the inequality*

$$\sum_n \lambda_n (\Lambda_n^*)^{-c} \left(\sum_{k=n}^{\infty} \lambda_k x_k \right)^p \le K \sum_n \lambda_n (\Lambda_n^*)^{p-c} x_n^p$$

holds with some $K > 0$ if and only if (Λ_n^*/λ_n) is bounded.

The proof follows similar lines as that of Theorems 11.1 and 12.1. A result cited in **BIII**, see (25) there, even allows to determine the best-possible constants K in the above inequalities. For example, in (b) it is

$$K = \sup_n \lambda_n^{1-1/p} \Lambda_n^{c/p-1} \left(\sum_{k=1}^{n} \lambda_k \Lambda_k^{-c} \right)^{1/p}.$$

13 Further Applications

In this section we discuss various further applications of the results in Chapters I and II. Most of them are concerned with representations of various sequence spaces to be found in the literature.

Equality of the spaces ces(p) and cop(p)

We have seen in the Introduction that Hardy's inequality, in its qualitative version, can be written as an inclusion between sequence spaces:

$$l^p \subset \text{ces}(p) \qquad (p > 1).$$

In order to do the same for the inequalities (9.5) and (10.1) of Copson and for Elliott's inequality (see [40, Theorems 338, 345]), Bennett has introduced the sequence spaces

$$\text{cop}(p) = \left\{ x : \sum_n \left(\sum_{k=n}^{\infty} \frac{|x_k|}{k} \right)^p < \infty \right\} \qquad \text{for} \quad p > 0,$$
$$\text{ell}(p) = \left\{ x : \sum_n \left(\frac{1}{n} \sum_{k=n}^{\infty} |x_k| \right)^p < \infty \right\} \qquad \text{for} \quad 0 < p < 1$$

(**BIV**, §§6, 16). The inequalities of Copson and Elliott now read

$$l^p \subset \text{cop}(p) \qquad (p > 1),$$
$$\text{cop}(p) \subset l^p \qquad (0 < p < 1),$$
$$\text{ell}(p) \subset l^p \qquad (0 < p < 1).$$

Of course, these four inclusion results do not imply any relation among the spaces ces(p), cop(p) and ell(p). But Bennett (**BIV**, pp. 26, 88) found as a by-product of his results the somewhat surprising fact that, indeed,

$$\text{cop}(p) = \text{ces}(p) \qquad \text{for} \quad p > 1$$

and

$$\text{cop}(p) = \text{ell}(p) \qquad \text{for} \quad 0 < p < 1.$$

We add that the case $p = 2$ was already noted by Hardy [35]. The missing case $p = 1$ with cop(1) = ell(1) = l^1 is a triviality.

Bennett's major concern is with the corresponding quantitative version of finding estimates between the norms involved, see **BIV** §§10 and 16. But as far as the qualitative aspect is concerned, his identities are special cases of two results of Askey and Boas [18, Lemma 6.18], [5, Lemma]. They show (some change of variables is required) that for $0 < p < \infty$ and real parameters α, β, α'

and β' with $\alpha < -1/p$ and $\alpha' > -1/p$ the conditions

$$\sum_n \left(n^{\alpha} \sum_{k=1}^{n} k^{\beta} |x_k|\right)^p < \infty$$

and

$$\sum_n \left(n^{\alpha'} \sum_{k=n}^{\infty} k^{\beta'} |x_k|\right)^p < \infty$$

are equivalent for all sequences $\boldsymbol{x}$ if $\alpha + \beta = \alpha' + \beta'$. This implies the equality of cop(p) with ces(p) for $p > 1$ and with ell(p) for $p < 1$. We shall now deduce the Askey-Boas theorem from the results in Chapter I. We consider more generally an arbitrary sequence $\boldsymbol{a}$ of non-negative terms with $a_1 > 0$ and $\sum_n a_n = \infty$. We set

$$A_n = \sum_{k=1}^{n} a_k$$

and assume that $A_{n+1} = O(A_n)$.

Theorem 13.1 *Let $0 < p, q < \infty$, and let α, β, α' and β' be real numbers with $\alpha < -p/q$ and $\alpha' > -p/q$. If $\alpha + \beta = \alpha' + \beta'$, then for any sequence $\boldsymbol{x}$ the conditions*

$$\sum_n a_{n+1} \left(A_n^{\alpha} \sum_{k=1}^{n} A_k^{\beta} |x_k|^p\right)^{q/p} < \infty$$

and

$$\sum_n a_n \left(A_n^{\alpha'} \sum_{k=n}^{\infty} A_k^{\beta'} |x_k|^p\right)^{q/p} < \infty$$

are equivalent.

Proof. We apply Theorem 2.1 with $s_n = A_n^{-1}$ and substitute $-(\alpha/p + \beta/p + 1/q) = -(\alpha'/p + \beta'/p + 1/q)$ for α. The result follows from the equivalence of (2.4ii) and (2.4iii) when we assign the parameters β, γ, δ there the values $-1, -1 - \alpha q/p, -\beta$ and $-1, -1 - \alpha' q/p, -\beta'$, respectively. □

The Askey-Boas theorem is the special case $\boldsymbol{a} = \boldsymbol{1}, p = 1$ of this result.

Multipliers between spaces of type c and d

It would be of great interest to determine all multipliers between any two sequence spaces of type $c(\boldsymbol{a},p,q)$ and $d(\boldsymbol{a},p,q)$. At present we are not able to do this since these spaces, in general, translate into weighted $l(\boldsymbol{m},p,q)$-spaces with different index sequences $\boldsymbol{m}$ so that Kellogg's theorem 7.1 is not applicable.

However, our method continues to work if the index sequences $\boldsymbol{m}$ happen to coincide or can be made to do so. We give an example here.

Example 13.2 In §14 of **BIV**, Bennett has determined the space of multipliers from any one of the spaces $\mathrm{cop}(p), d(p), g(p)$ and l^p into any one of the spaces $\mathrm{cop}(q), d(q), g(q)$ and l^q with $0 < p, q \le \infty$, where

$$\mathrm{cop}(p) = \left\{ \boldsymbol{x} : \sum_n \left(\sum_{k=n}^{\infty} \frac{|x_k|}{k} \right)^p < \infty \right\},$$

$$d(p) = \left\{ \boldsymbol{x} : \sum_n \sup_{k \ge n} |x_k|^p < \infty \right\},$$

$$g(p) = \left\{ \boldsymbol{x} : \sup_n \frac{1}{n} \sum_{k=1}^{n} |x_k|^p < \infty \right\}.$$

If we apply Theorems 3.1 and 3.2 to these sequence spaces, a complication arises. For $p < \infty$ the representations in block form we obtain have blocks $[2^{p\nu}, 2^{p(\nu+1)})$, and these vary with p. To remedy this situation we recall a remark in Section 4 that the base 2 can be replaced by any other base $\rho > 1$ in Chapter I. Choosing $\rho = 2^{1/p}$, we can now deduce that

$$\mathrm{cop}(p) = \left\{ \boldsymbol{x} : \sum_\nu 2^{\nu(1-p)} \left(\sum_{k \in D_\nu} |x_k| \right)^p < \infty \right\},$$

$$d(p) = \left\{ \boldsymbol{x} : \sum_\nu 2^\nu \sup_{k \in D_\nu} |x_k|^p < \infty \right\},$$

$$g(p) = \left\{ \boldsymbol{x} : \sup_\nu 2^{-\nu/p} \left(\sum_{k \in D_\nu} |x_k|^p \right)^{1/p} < \infty \right\},$$

and clearly

$$l^p = \left\{ \boldsymbol{x} : \sum_\nu \left(\sum_{k \in D_\nu} |x_k|^p \right)^{p/p} < \infty \right\},$$

where $D_\nu = [2^\nu, 2^{\nu+1})$ stands for the dyadic blocks, with certain modifications for $p = \infty$. We have thus achieved that the blocks are the same for all spaces involved, so that Kellogg's theorem and the techniques of Section 7 enable us to calculate the desired multiplier spaces and thus confirm the tables on pp. 69 - 70 of **BIV** (with $I(\alpha, \beta)$ as in **BIV**, Proposition 15.4). We remark that in constrast to Bennett's ingenious juggling with various suitable methods, the results of Chapter I provide a unified and elementary technique that reduces the task of determining the multiplier spaces to simple, if tedious, calculations.

In addition, our calculations have shown that Bennett's tables have to be taken with a pinch of salt. Indeed, the spaces $\mathrm{cop}(\infty)$ and $H(\infty, \beta)$ with $\beta < \infty$ need to be understood as

$$\mathrm{ces}(\infty) = \left\{ x : \sup_n \frac{1}{n} \sum_{k=1}^{n} |x_k| < \infty \right\}$$

and

$$g(\beta) = \left\{ x : \sup_n \frac{1}{n} \sum_{k=1}^{n} |x_k|^\beta < \infty \right\},$$

respectively, contrary to what one would expect as limiting cases of $\mathrm{cop}(p)$ as $p \to \infty$ and $H(\alpha, \beta)$ as $\alpha \to \infty$ (but see **BIV**, p. 82 for H). This is of relevance, for instance, in Item 24 for $p = q$ (compare also Item 8).

We remark that the space $G(\infty, \beta)$ with $\beta < \infty$, where a similar problem might have arisen, does not appear in the tables, nor does $I(\infty, \beta)$ with $\beta < \infty$.

Spaces of strongly Cesàro summable sequences

A sequence x is called *strongly Cesàro summable with index* $p > 0$ if there is some number l with

$$\frac{1}{n} \sum_{k=1}^{n} |x_k - l|^p \to 0$$

as $n \to \infty$. The space of all such sequences is denoted by w_p. In 1965, D. Borwein [21] succeeded in determining the dual of w_p for $1 \leq p < \infty$. His derivation is based on an application of the blocking technique. In a similar fashion, Maddox [73] gave the dual of w_p for $0 < p < 1$. In order to state the Borwein-Maddox result we note that

$$w_p = w_p^0 \oplus < \mathbf{1} >,$$

where

$$w_p^0 = \left\{ x : \frac{1}{n} \sum_{k=1}^{n} |x_k|^p \to 0 \right\}.$$

Since w_p^0, in its natural quasi-norm, is a sequence space with the AK-property, it now suffices to find the Köthe dual of w_p^0 to determine with it the dual of w_p, compare the beginning of Section 7. According to Borwein and Maddox this Köthe dual turns out to be

$$(w_p^0)^\times = \left\{ x : \sum_{\nu} 2^{\nu/p} \Big(\sum_{k \in D_\nu} |x_k|^{p^\times} \Big)^{1/p^\times} < \infty \right\},$$

where $p^\times = p \to 1$ and $D_\nu = [2^\nu, 2^{\nu+1})$ are the dyadic blocks. Both authors stop with this representation, which is not quite satisfactory as the original problem was stated in terms of section norms and the solution is given in terms of block norms. Thus a re-translation is needed as provided by Section 2.

Theorem 13.3 *Let* $0 < p < \infty$. *Then*

$$\begin{aligned} (w_p^0)^\times &= \left\{ x : \sum_{n} \Big(\frac{1}{n} \sum_{k=n}^{\infty} |x_k|^{p^\times} \Big)^{1/p^\times} < \infty \right\} \qquad &\text{if } p > 1, \\ &= \left\{ x : \sum_{n} \Big(\frac{1}{n} \Big)^{1-1/p} \sup_{k \ge n} |x_k| < \infty \right\} \qquad &\text{if } p \le 1. \end{aligned}$$

The equivalent representation $(w_p^0)^\times = \{x : \sum_n \big(\sum_{k=n}^{\infty} 1/k|x_k|^{p^\times}\big)^{1/p^\times} < \infty\}$ for $p > 1$ (cf. Theorem 13.1) and the above representation for $p = 1$ have already been obtained by Balser, Jurkat and Peyerimhoff [7] in a direct way without taking recourse to Borwein's result. For $p = 1$ the above representation was also derived by Belinskii, Liflyand and Trigub [10, Lemma 1] who in fact show that it yields the dual norm in this case.

Balser, Jurkat and Peyerimhoff have moreover determined the dual of the space of strongly summable sequences with respect to any positive weighted mean method, giving the dual in section form, as is appropriate. On the other hand, there is an extended literature on these and more general spaces that uses the blocking technique following Borwein and Maddox. In many cases the results are stated in block form so that a re-translation into section form is possible with the aid of Chapter I.

A theorem of Fefferman

Let $\boldsymbol{a}$ be any non-negative sequence. The entries of the *Hankel matrix* $H_{\boldsymbol{a}}$ are given by

$$h_{nk} = a_{n+k}$$

for $n, k \geq 0$. We shall consider here sequences $\boldsymbol{x}$ with indices starting from 0. Then, according to an unpublished result of C. Fefferman (see [4, p. 264]), the Hankel matrix $H_{\boldsymbol{a}}$ maps l^2 into l^2 if and only if

$$\sup_{N\geq 1} \sum_{\nu=1}^{\infty} \left(\sum_{k=\nu N}^{(\nu+1)N-1} a_k \right)^2 < \infty. \tag{13.1}$$

We refer to [20], [6, p. 41] and **BI**, p. 423 for further information. Fefferman's condition differs from the norms in block form that we have considered so far in that we have here a supremum over countably many such norms. Taking account of the remark on constants in Section 4 we may still apply the results of Section 2 to obtain the following.

Proposition 13.4 *For any non-negative sequence* $\boldsymbol{a}$*, Fefferman's condition* (13.1) *is equivalent to*

$$\sup_{N\geq 1} \frac{1}{N} \sum_{n=N}^{\infty} \left(\sum_{k=0}^{\infty} a_{n+k} r_N^k \right)^2 < \infty, \tag{13.2}$$

where $r_N = 2^{-1/N}$.

Proof. We fix $N \geq 1$ and set $s_n = 2^{-n/N}$ for $n \geq 0, m_\nu = \nu N$ for $\nu \geq 0$, $\alpha = 0, \beta = \gamma = -2$ and $\delta = 1$. Then the sequences (s_n) and (m_ν) are correlated in the sense of Section 2. Applying Theorem 2.1 to the sequence $\boldsymbol{x} = (a_{N+n})_n$ we see that

$$\rho_N := \sum_{\nu=1}^{\infty} \left(\sum_{k=\nu N}^{(\nu+1)N-1} a_k \right)^2 < \infty$$

is equivalent to

$$\sigma_N := (1 - r_N^2) \sum_{n=N}^{\infty} \left(\sum_{k=0}^{\infty} a_{n+k} r_N^k \right)^2 < \infty.$$

Now, since $s_n \leq 1$ for all n independently of N, a remark in Section 4 shows that the constants relating ρ_N and σ_N for $N \in \mathbb{N}$ can be chosen to be independent of N. Since $1 - r_N^2 \sim 1/N$, the equivalence of (13.1) and (13.2) follows. □

The proposition implies a result of Bonsall [20, Remark 3]. He shows that the following variant of (13.2) also characterises when $H_{\boldsymbol{a}}$ maps l^2 into l^2:

$$\sup_{r\in[0,1)} (1-r^2)\sum_{n=0}^{\infty}\left(\sum_{k=0}^{\infty} a_{n+k}r^k\right)^2 < \infty.$$

Since this can also be written as

$$\sup_{r\in[0,1)} \frac{\|H_{\boldsymbol{a}}(r^n)\|_2}{\|(r^n)\|_2} < \infty,$$

it is clearly a necessary condition. On the other hand, since $1/N \sim 1-r_N^2$, its sufficiency is immediate from Proposition 13.4.

Integrability classes for trigonometric series

The Fourier expansion of an even function $f \in L^1[-\pi,\pi]$ is given by a cosine series

$$\frac{a_0}{2} + \sum_{n=1}^{\infty} a_n \cos nx.$$

No description of the space $\widehat{L_1}$ of all sequences $\boldsymbol{a}$ arising in this way is known when we ask for a description purely in terms of sequences. This has led to a search for ever bigger sequence spaces E that are contained in $\widehat{L_1}$. Such sequence spaces are called *integrability classes for cosine series.* We refer to [26] and [6] for good surveys on the subject.

Norms in block form were introduced into this area by Fomin [32]. Recently, Buntinas and Tanović-Miller [25, Theorem 5] and Aubertin and Fournier [6, Theorem 1] have independently found one of the largest integrability classes to date. It consists of all sequences $\boldsymbol{a} \in c_0$ that satisfy

$$\sum_{N=1}^{\infty} \frac{1}{N}\left[\sum_{\nu=1}^{\infty}\left(\sum_{k=\nu N}^{(\nu+1)N-1} |\Delta a_k|\right)^2\right]^{1/2} < \infty, \tag{13.3}$$

where $\Delta a_k = a_k - a_{k+1}$. This condition is obviously closely related to Fefferman's condition (13.1). By the proof of Proposition 13.4 we obtain the following equivalent description.

Proposition 13.5 *For any sequence* $\boldsymbol{a}$, *condition* (13.3) *is equivalent to*

$$\sum_{N=1}^{\infty}\left[\frac{1}{N^3}\sum_{n=N}^{\infty}\left(\sum_{k=0}^{\infty} |\Delta a_{n+k}| r_N^k\right)^2\right]^{1/2} < \infty,$$

where $r_N = 2^{-1/N}$.

Chapter IV

Integral Analogues

14 Introduction

In its integral form, Hardy's inequality asserts that for $p > 1$ there is a constant $K > 0$ such that

$$\int_0^\infty \left(\frac{1}{x}\int_0^x f(t)dt\right)^p dx \leq K \int_0^\infty f(x)^p dx$$

holds for every non-negative measurable function f on $[0, \infty)$ ([37], see also [36]). The best-possible constant turned out to be $K = (\frac{p}{p-1})^p$, as in the discrete case.

The history of Hardy's inequality falls largely into two parts. In a classical period leading up to the Hardy-Littlewood-Pólya treatise [40] there seemed to be an emphasis on elementary inequalities, that is, discrete inequalities involving (finitely many) non-negative variables. Nonetheless most authors studied their integral analogues alongside them. Interest in Hardy's and related inequalities revived in the 1960's, but since then the discrete and the integral versions have been studied more or less separately, and there is a considerably larger body of research on the integral form with Hardy's inequality being generalised in various directions. For a survey we refer to [81], see also [76].

A great deal of research is devoted to Hardy's inequalities with (non-negative) weights

$$(14.1) \qquad \left(\int_a^b \left(u(x)\int_a^x f(t)dt\right)^q dx\right)^{1/q} \leq K \left(\int_a^b (v(x)f(x))^p dx\right)^{1/p}$$

and

$$(14.2)\qquad \left(\int_a^b \left(u(x)\int_x^b f(t)dt\right)^q dx\right)^{1/q} \le K\left(\int_a^b (v(x)f(x))^p dx\right)^{1/p},$$

that is, the integral analogues of inequalities (9.1) and (9.4). We remark that it is purely for historical reasons that v appears here on the right-hand side rather than on the left-hand side. The weights u and v for which (14.1) and (14.2) hold for all non-negative f have been completely characterised. The solution, in the various cases of p and q, is connected with the names of M. Artola, G. Talenti, G. Tomaselli, B. Muckenhoupt, V. G. Maz'ja, A. L. Rosin, J. S. Bradley, E. Sawyer, G. Sinnamon, and others. The characterising conditions on u and v are analogues of the ones obtained by Bennett (**BIII**) for the discrete case, with one notable exception: In the case of $0 < p < 1$ only trivial weights lead to a valid inequality, see [86, Theorem 2].

The corresponding lower boundedness problem, that is, the set of inequalities

$$(14.3)\qquad \left(\int_a^b \left(u(x)\int_a^x f(t)dt\right)^q dx\right)^{1/q} \ge K\left(\int_a^b (v(x)f(x))^p dx\right)^{1/p}$$

and

$$(14.4)\qquad \left(\int_a^b \left(u(x)\int_x^b f(t)dt\right)^q dx\right)^{1/q} \ge K\left(\int_a^b (v(x)f(x))^p dx\right)^{1/p}$$

with some $K > 0$ has received much less attention; these are Copson's inequalities (10.2) and (10.4) in integral form. This problem has only been studied by Beesack and Heinig [9] and Bennett (see **BIV**, p. 37), and it is only solved for $0 < q \le p < 1$. Bennett has asked for a characterisation in the case $0 < p < q < 1$.

In this chapter we briefly outline the integral analogue of the blocking technique of Chapter I which then enables us to treat the problems considered in Chapters II and III in their integral form. In particular, one obtains a new approach to the weighted Hardy inequalities (14.1) and (14.2), and we can solve the corresponding lower boundedness problems (14.3) and (14.4) for all p and q, answering Bennett's question. As always, our results are qualitative, without estimates for the constants K involved.

Throughout the remainder of this chapter we assume that (a,b) is a fixed interval with $-\infty \le a < b \le \infty$. It suffices to consider open intervals only.

15 The Blocking Technique

Let u be a non-negative measurable function on (a, b) and let $0 < p, q \leq \infty$. For any measurable function f on (a, b) we consider the expressions in section form

$$\int_a^b \left[u(x) \left(\int_a^x |f(t)|^p \, dt \right)^{1/p} \right]^q dx$$

and

$$\int_a^b \left[u(x) \left(\int_x^b |f(t)|^p \, dt \right)^{1/p} \right]^q dx,$$

which we want to replace by expressions in block form. The situation we face here is more symmetric than in the discrete case in that now both end-points a and b are possibly critical. In order to capture this we need to partition (a, b) into a doubly infinite sequence of intervals I_ν with end-points t_ν and $t_{\nu+1}$ $(\nu \in \mathbb{Z})$, where $a \leq t_\nu \leq t_{\nu+1} \leq b$, $\lim_{\nu \to -\infty} t_\nu = a$ and $\lim_{\nu \to \infty} t_\nu = b$. Thus the corresponding expressions in block form are now

$$\sum_{\nu \in \mathbb{Z}} \left[\frac{1}{2^{\nu\alpha}} \left(\int_{I_\nu} |f(t)|^p \, dt \right)^{1/p} \right]^q,$$

where $\alpha \in \mathbb{R}$. A partition (I_ν) (or the corresponding sequence $\boldsymbol{t} = (t_\nu)$) on the one hand and a positive non-increasing function s on (a, b) on the other hand will be called *correlated* if

$$\frac{1}{2^\nu} \geq s(x) \geq \frac{1}{2^{\nu+1}} \quad \text{for } t_\nu < x < t_{\nu+1} \quad (\nu \in \mathbb{Z})$$

holds. We are more liberal here than in the discrete case. Also, we do not require that $\lim_{x \to b} s(x) = 0$.

We first transform again from block form into section form. Some conditions in section form that we have obtained are in terms of Lebesgue-Stieltjes integrals. In the sequel, we say that f is L^p at a if $f \in L^p(a, x)$ for some $x > a$, and similarly for the right end-point.

We fix a partition $\boldsymbol{t}$ and a function s that is correlated to it.

Theorem 15.1 *Let $\alpha \in \mathbb{R}$, and let f be any measurable function on (a, b).*
(a) *Let $0 < p, q < \infty$. Then the condition*

$$\sum_{\nu \in \mathbb{Z}} \left[\frac{1}{2^{\nu\alpha}} \left(\int_{I_\nu} |f(t)|^p \, dt \right)^{1/p} \right]^q < \infty$$

is equivalent to any of the following conditions, where $\gamma \neq 0$ and δ are real numbers with $\gamma/q + \delta/p = \alpha$:

$$\int_a^b \left(\int_a^x s(t)^\delta |f(t)|^p \, dt \right)^{q/p} d(-s(x)^\gamma) < \infty$$
$$\text{and, in addition, } f \text{ is } L^p \text{ at } b \text{ if } \lim_{x \to b} s(x) \neq 0, \qquad \text{if} \quad \gamma > 0,$$

$$\int_a^b \left(\int_x^b s(t)^\delta |f(t)|^p \, dt \right)^{q/p} d(s(x)^\gamma) < \infty$$
$$\text{and, in addition, } f \text{ is } L^p \text{ at } a \text{ if } \lim_{x \to a} s(x) \neq \infty, \qquad \text{if} \quad \gamma < 0,$$

$$\int_a^b s(x)^{\gamma+\delta} |f(x)|^p \left(\int_a^x s(t)^\delta |f(t)|^p \, dt \right)^{q/p - 1} dx < \infty \qquad \text{if} \quad \gamma > 0,$$

$$\int_a^b s(x)^{\gamma+\delta} |f(x)|^p \left(\int_x^b s(t)^\delta |f(t)|^p \, dt \right)^{q/p - 1} dx < \infty \qquad \text{if} \quad \gamma < 0.$$

(b) *Let $0 < q < \infty$. Then the condition*

$$\sum_{\nu \in \mathbb{Z}} \left(\frac{1}{2^{\nu\alpha}} \operatorname*{ess\,sup}_{I_\nu} |f(t)| \right)^q < \infty$$

is equivalent to any of the following conditions, where $\gamma \neq 0$ and δ are real numbers with $\gamma/q + \delta = \alpha$:

$$\int_a^b \left(\operatorname*{ess\,sup}_{t \le x} s(t)^\delta |f(t)| \right)^q d(-s(x)^\gamma) < \infty$$
$$\text{and, in addition, } f \text{ is } L^\infty \text{ at } b \text{ if } \lim_{x \to b} s(x) \neq 0, \qquad \text{if} \quad \gamma > 0,$$

$$\int_a^b \left(\operatorname*{ess\,sup}_{t \ge x} s(t)^\delta |f(t)| \right)^q d(s(x)^\gamma) < \infty$$
$$\text{and, in addition, } f \text{ is } L^\infty \text{ at } a \text{ if } \lim_{x \to a} s(x) \neq \infty, \qquad \text{if} \quad \gamma < 0.$$

(c) *Let $0 < p < \infty$. Then the condition*

$$\sup_{\nu \in \mathbb{Z}} \frac{1}{2^{\nu\alpha}} \left(\int_{I_\nu} |f(t)|^p \, dt \right)^{1/p} < \infty$$

is equivalent to any of the following conditions, where $\gamma \neq 0, \rho > 0, \sigma > 0$

and δ are real numbers with $\gamma + \delta/p = \alpha$:

$$\operatorname*{ess\,sup}_{x\in(a,b)} s(x)^\gamma \left(\int_a^x s(t)^\delta |f(t)|^p\,dt\right)^{1/p} < \infty \qquad \text{if } \gamma > 0,$$

$$\operatorname*{ess\,sup}_{x\in(a,b)} s(x)^\gamma \left(\int_x^b s(t)^\delta |f(t)|^p\,dt\right)^{1/p} < \infty \qquad \text{if } \gamma < 0,$$

$$\int_a^x \left(\int_a^t s(\tau)^\delta |f(\tau)|^p\,d\tau\right)^{(\rho+\sigma)/p} d(-s(t)^{\rho\gamma}) = O\left(\int_a^x s(t)^\delta |f(t)|^p\,dt\right)^{\sigma/p}$$

as $x \to b$, and, in addition, f is L^p at b if $\lim_{x\to b} s(x) \neq 0$, if $\gamma > 0$,

$$\int_x^b \left(\int_t^b s(\tau)^\delta |f(\tau)|^p\,d\tau\right)^{(\rho+\sigma)/p} d(s(t)^{\rho\gamma}) = O\left(\int_x^b s(t)^\delta |f(t)|^p\,dt\right)^{\sigma/p}$$

as $x \to a$, and, in addition, f is L^p at a if $\lim_{x\to a} s(x) \neq \infty$, if $\gamma < 0$.

(d) *The condition*

$$\sup_{\nu\in\mathbb{Z}} \frac{1}{2^{\nu\alpha}} \operatorname*{ess\,sup}_{t\in I_\nu} |f(t)| < \infty$$

is equivalent to the following conditions, where γ and δ are real numbers with $\gamma + \delta = \alpha$:

$$\operatorname*{ess\,sup}_{x\in(a,b)} s(x)^\gamma \operatorname*{ess\,sup}_{t\le x} s(t)^\delta |f(t)| < \infty \qquad \text{if } \gamma \ge 0,$$

$$\operatorname*{ess\,sup}_{x\in(a,b)} s(x)^\gamma \operatorname*{ess\,sup}_{t\ge x} s(t)^\delta |f(t)| < \infty \qquad \text{if } \gamma \le 0,$$

$$\operatorname*{ess\,sup}_{x\in(a,b)} s(x)^\alpha |f(x)| < \infty.$$

The proof is similar to that in the discrete case. We note that as a consequence of the symmetry of the present situation the results for $\gamma > 0$ and $\gamma < 0$ are equivalent. This can be derived by the substitution $x \mapsto (b+a) - x$.

Remark 15.2 (i) If a condition contains an integral $\int_a^x$ or $\int_x^b$, this condition is understood to imply the finiteness of these integrals for all $x < b$ and $x > a$, respectively. A similar remark applies to suprema. Also, in 15.1(a) an expression $0 \cdot 0^r$ with $r < 0$ may occur; it is to be taken as 0.

(ii) In most applications the function s will be absolutely continuous on every compact subinterval of (a,b). In that case the measure $d(\pm s(x)^\beta)$ appearing in some conditions coincides with $(\pm s(x)^\beta)' dx$.

We turn to the transformation from section form into block form. Let now u be any non-negative measurable function on (a,b).

Theorem 15.3 *Let f be a measurable function on (a,b).*

(a) *Let $0 < p, q < \infty$. Assume that $0 < \int_t^b u(x)^q dx < \infty$ for all $t \in (a,b)$, and define t_ν as the least t such that $(\int_t^b u(x)^q dx)^{1/q} \leq 1/2^\nu \quad (\nu \in \mathbb{Z})$. Then*

$$\int_a^b \left[u(x) \left(\int_a^x |f(t)|^p dt \right)^{1/p} \right]^q dx < \infty$$

is equivalent to

$$\sum_{\nu \in \mathbb{Z}} \left[\frac{1}{2^\nu} \left(\int_{I_\nu} |f(t)|^p dt \right)^{1/p} \right]^q < \infty.$$

(b) *Let $0 < q < \infty$. Assume that $0 < \int_t^b u(x)^q dx < \infty$ for all $t \in (a,b)$, and define a partition $\mathbf{t}$ as in* (a). *Then*

$$\int_a^b \left(u(x) \operatorname*{ess\,sup}_{t \leq x} |f(t)| \right)^q dx < \infty$$

is equivalent to

$$\sum_{\nu \in \mathbb{Z}} \left(\frac{1}{2^\nu} \operatorname*{ess\,sup}_{t \in I_\nu} |f(t)| \right)^q < \infty.$$

(c) *Let $0 < p < \infty$. Assume that $0 < \operatorname{ess\,sup}_{x \geq t} u(x) < \infty$ for all $t \in (a,b)$, and define t_ν as the least t such that $\operatorname{ess\,sup}_{x \geq t} u(x) \leq 1/2^\nu$ $(\nu \in \mathbb{Z})$, setting $t_\nu = b$ if no such t exists. Then*

$$\operatorname*{ess\,sup}_{x \in (a,b)} u(x) \left(\int_a^x |f(t)|^p dt \right)^{1/p} < \infty$$

is equivalent to

$$\sup_{\nu \in \mathbb{Z}} \frac{1}{2^\nu} \left(\int_{I_\nu} |f(t)|^p dt \right)^{1/p} < \infty.$$

(d) *Assume that* $0 < \operatorname{ess\,sup}_{x \geq t} u(x) < \infty$ *for all* $t \in (a,b)$, *and define a partition* $\mathbf{t}$ *as in* (c). *Then*

$$\operatorname*{ess\,sup}_{x\in(a,b)} u(x) \operatorname*{ess\,sup}_{t\leq x} |f(t)| < \infty$$

is equivalent to

$$\sup_{\nu\in\mathbb{Z}} \frac{1}{2^\nu} \operatorname*{ess\,sup}_{t\in I_\nu} |f(t)| < \infty.$$

Theorem 15.4 *Let* f *be a measurable function on* (a,b).

(a) *Let* $0 < p,q < \infty$. *Assume that* $0 < \int_a^t u(x)^q dx < \infty$ *for all* $t \in (a,b)$, *and define* t_ν *as the least* t *such that* $(\int_a^t u(x)^q dx)^{1/q} \geq 2^\nu$ $(\nu \in \mathbb{Z})$, *setting* $t_\nu = b$ *if no such* t *exists. Then*

$$\int_a^b \left[u(x) \left(\int_x^b |f(t)|^p dt \right)^{1/p} \right]^q dx < \infty$$

is equivalent to

$$\sum_{\nu\in\mathbb{Z}} \left[2^\nu \left(\int_{I_\nu} |f(t)|^p dt \right)^{1/p} \right]^q < \infty.$$

(b) *Let* $0 < q < \infty$. *Assume that* $0 < \int_a^t u(x)^q dx < \infty$ *for all* $t \in (a,b)$, *and define a partition* $\mathbf{t}$ *as in* (a). *Then*

$$\int_a^b \left(u(x) \operatorname*{ess\,sup}_{t\geq x} |f(t)| \right)^q dx < \infty$$

is equivalent to

$$\sum_{\nu\in\mathbb{Z}} \left(2^\nu \operatorname*{ess\,sup}_{t\in I_\nu} |f(t)| \right)^q < \infty.$$

(c) *Let* $0 < p < \infty$. *Assume that* $0 < \operatorname{ess\,sup}_{x\leq t} u(x) < \infty$ *for all* $t \in (a,b)$, *and define* t_ν *as the infimum of all* t *such that* $\operatorname{ess\,sup}_{x\leq t} u(x) \geq 2^\nu$ $(\nu \in \mathbb{Z})$, *setting* $t_\nu = b$ *if no such* t *exists. Then*

$$\operatorname*{ess\,sup}_{x\in(a,b)} u(x) \left(\int_x^b |f(t)|^p dt \right)^{1/p} < \infty$$

is equivalent to

$$\sup_{\nu\in\mathbb{Z}} 2^\nu \left(\int_{I_\nu} |f(t)|^p dt \right)^{1/p} < \infty.$$

(d) *Assume that* $0 < \operatorname{ess\,sup}_{x \leq t} u(x) < \infty$ *for all* $t \in (a,b)$, *and define a partition* $\boldsymbol{t}$ *as in* (c). *Then*

$$\operatorname*{ess\,sup}_{x\in(a,b)} u(x) \operatorname*{ess\,sup}_{t\geq x} |f(t)| < \infty$$

is equivalent to

$$\sup_{\nu\in\mathbb{Z}} 2^{\nu} \operatorname*{ess\,sup}_{t\in I_\nu} |f(t)| < \infty.$$

The last two results are equivalent via the substitution $x \mapsto (b+a) - x$.

16 The Function Spaces $C(u,p,q), D(u,p,q)$ and $(L^p, l^q, \boldsymbol{t})$

The results of the previous section suggest the introduction of corresponding function spaces. We let f always denote measurable functions on the interval (a,b). For any non-negative measurable function u on (a,b) and $0 < p,q \leq \infty$ we define

$$C(u,p,q) = \left\{ f : \int_a^b \left[u(x) \left(\int_a^x |f(t)|^p dt \right)^{1/p} \right]^q < \infty \right\}$$

and

$$D(u,p,q) = \left\{ f : \int_a^b \left[u(x) \left(\int_x^b |f(t)|^p dt \right)^{1/p} \right]^q < \infty \right\},$$

and for any doubly infinite sequence $\boldsymbol{t} = (t_\nu)_{\nu\in\mathbb{Z}}$ with $a \leq t_\nu \leq t_{\nu+1} \leq b, \lim_{\nu\to-\infty} t_\nu = a$ and $\lim_{\nu\to\infty} t_\nu = b$ we set

$$(L^p, l^q, \boldsymbol{t}) = \left\{ f : \sum_{\nu\in\mathbb{Z}} \left(\int_{t_\nu}^{t_{\nu+1}} |f(t)|^p dt \right)^{q/p} < \infty \right\},$$

with the usual modifications if p or q is infinite.

Many function spaces from the literature, in particular from Harmonic Analysis, are covered by the spaces C and D. Let us only mention the Beurling algebras A^p and A^*, see [34], [47], [10]. In connection with Hardy's inequality the *Cesàro function space*

$$\mathrm{Ces}(p) = \left\{ f : \int_0^\infty \left(\frac{1}{x} \int_0^x |f(t)| dt \right)^p dx < \infty \right\},$$

where $1 < p \leq \infty$, is of interest. It has been introduced in [101] and subsequently studied in [56], [91] and **BIV**, §21. Further examples can be found in Section 18.

The function spaces C and D can be studied in analogy to the investigations in Chapter II, based on the transformation rules of Section 15. We shall not pursue this here.

The spaces $(L^p, l^q, \boldsymbol{t})$ are, of course, well known. They were introduced by F. Holland [43] in the special case of $(a, b) = \mathbb{R}$ and $t_\nu = \nu$ following an idea of Wiener. The spaces are called *amalgams* (of L^p and l^q), more precisely *stretched* or *irregular* amalgams in case of general $\boldsymbol{t}$. We refer to [33] for a thorough survey, see also [30], [31]. At this point we only state the characterisation of (pointwise) multipliers between these spaces. It is not difficult to see that for $0 < p, q, r, s \leq \infty$ we have

$$(16.1) \qquad \mathcal{M}((L^p, l^q, \boldsymbol{t}), (L^r, l^s, \boldsymbol{t})) = (L^{p \to r}, l^{q \to s}, \boldsymbol{t}) \qquad \text{if} \quad p \geq r$$
$$= \{0\} \qquad \text{if} \quad p < r.$$

The triviality in the case $p < r$, which, incidentally, seems to be at odds with a result in [43, p. 298], is due to the fact that no function $f \neq 0$ can be a multiplier form $L^p(I)$ into $L^r(I)$ for $p < r$ and any interval I.

17 The Inequalities of Hardy and Copson with Weights

The inequalities (14.1) - (14.4) can be treated in analogy to the investigation in Chapter III. Everything reduces to an application of the multiplier result (16.1) together with the appropriate transformations back and forth according to Section 15 – at least for 'non-trivial' u and v.

In the case of Hardy's inequality with weights we are thus able to give new and elementary proofs of the known characterisations. In particular, (16.1) explains why only trivial weights are possible in the case $p < 1$.

We state here the result for Copson's inequality with weights in the case of $p < 1, p < q < \infty$. This completes the investigation of Beesack and Heinig [9] and answers the question of Bennett (**BIV**, p. 37).

Theorem 17.1 *Let $0 < p < 1, p < q < \infty$, and let u and v be non-negative measurable functions on (a, b) such that*

$$0 < \int_t^b u(x)^q dx < \infty \quad \textit{for all} \quad t \in (a, b).$$

Then the weighted Copson inequality

$$\left(\int_a^b \left(u(x) \int_a^x f(t)dt\right)^q dx\right)^{1/q} \geq K \left(\int_a^b (v(x)f(x))^p dx\right)^{1/p} \tag{17.1}$$

holds for all non-negative measurable functions f if and only if any of the following equivalent conditions is satisfied:

$$\int_a^b u(x)^q \left(\int_x^b u(t)^q dt\right)^{-r/p} \left(\int_x^b v(t)^{-p^*} dt\right)^{-r/p^*} dx < \infty$$

and, in addition, $v \in L^{-p^*}(a, b)$ *if* $u \in L^q(a, b)$,

$$\int_a^b v(x)^{-p^*} \left(\int_x^b v(t)^{-p^*} dt\right)^{-r/q^*} \left(\int_x^b u(t)^q dt\right)^{-r/q} dx < \infty,$$

$$\int_a^b u(x)^q \left(\int_a^x \left[v(t)\left(\int_t^b u(\tau)^q d\tau\right)^{-1/p}\right]^{-p^*} dt\right)^{-r/p^*} dx < \infty,$$

where $\frac{1}{r} = \frac{1}{p} - \frac{1}{q}$.

We stress that the conditions have to be interpreted in accordance with Remark 15.2(i).

The initial assumption on u can easily be dispensed with. If u is an arbitrary non-negative measurable function on (a, b), then we set

$$a_0 = \sup \left\{\alpha : \int_\alpha^b u(x)^q dx = \infty\right\}$$

and

$$b_0 = \inf \left\{\beta : \int_\beta^b u(x)^q dx = 0\right\} = \inf \left\{\beta : u|_{(\beta,b)} = 0 \quad \text{a.e.}\right\}.$$

Then Copson's inequality (17.1) holds if and only if $v|_{(b_0,b)} = 0$ a.e. and any of the three equivalent conditions in the theorem holds for a_0 and b_0 replacing a and b.

Next we give the conditions characterising the validity of Copson's inequality in the remaining cases of p and q. We remark that the conditions for $0 < q \leq p < 1$ are due to Bennett (**BIV**, p. 37, see also [9]). As in Theorem 17.1 we assume that u and v are non-negative measurable functions on (a, b) such that, for all $t \in (a, b)$, we have

$$0 < \int_t^b u(x)^q dx < \infty \qquad \text{if} \quad q < \infty$$

and

$$0 < \sup_{x \geq t} u(x) < \infty \qquad \text{if} \quad q = \infty.$$

Then, in each case, each of the following equivalent conditions characterises when the weighted Copson inequality (17.1) holds for all non-negative measurable functions f:

(i) for $0 < q \leq p < 1$:

$$\operatorname*{ess\,sup}_{x \in (a,b)} \left(\int_x^b u(t)^q dt \right)^{-1/q} \left(\int_x^b v(t)^{-p^*} dt \right)^{-1/p^*} < \infty,$$

$$\operatorname*{ess\,sup}_{x \in (a,b)} \left(\int_x^b u(t)^q dt \right)^{1/q^*} \left(\int_x^b \left[v(t) \left(\int_t^b u(\tau)^q d\tau \right)^{-1} \right]^{-p^*} dt \right)^{-1/p^*} < \infty;$$

(ii) for $0 < p < 1, p < q < \infty$ see Theorem 17.1;

(iii) for $0 < p < 1, q = \infty$:

$$\int_a^b \left(\int_x^b v(t)^{-p^*} dt \right)^{-p/p^*} d((\operatorname*{ess\,sup}_{t \geq x} u(t))^{-p}) < \infty$$

and, in addition, $v \in L^{-p^*}(a, b)$ if $u \in L^\infty(a, b)$,

$$\int_a^b (\operatorname*{ess\,sup}_{t \geq x} u(t))^{-p} v(x)^{-p^*} \left(\int_x^b v(t)^{-p^*} dt \right)^{-p} dx < \infty;$$

(iv) for $0 < q \leq p = 1$:

$$\operatorname*{ess\,sup}_{x \in (a,b)} v(x) \left(\int_x^b u(t)^q dt \right)^{-1/q} < \infty;$$

(v) for $1 = p < q < \infty$:

$$\int_a^b u(x)^q \left(\int_x^b u(t)^q dt\right)^{-q^*} \operatorname*{ess\,sup}_{t \geq x} v(t)^{q^*} dx < \infty$$

and, in addition, $v \in L^\infty(a,b)$ if $u \in L^q(a,b)$,

$$\int_a^b u(x)^q \operatorname*{ess\,sup}_{t \leq x} \left(v(t)\left(\int_t^b u(\tau)^q d\tau\right)^{-1}\right)^{q^*} dx < \infty;$$

(vi) for $1 = p < q = \infty$:

$$\int_a^b \big(\operatorname*{ess\,sup}_{t \geq x} v(t)\big) d\big((\operatorname*{ess\,sup}_{t \geq x} u(t))^{-1}\big) < \infty$$

and, in addition, $v \in L^\infty(a,b)$ if $u \in L^\infty(a,b)$;

(vii) for $p > 1$: $v(x) = 0$ a.e.

This completes our discussion of the weighted Copson inequality (17.1) = (14.3). Its companion inequality (14.4) is obtained from (17.1) by the simple substitution $x \mapsto (b + a) - x$, from which a characterisation for this inequality follows.

18 Further Applications

In this section we collect some further applications of the integral version of the blocking technique.

Equality of the spaces Ces(p) and Cop(p)

In **BIV**, Theorem 21.1, Bennett observes that the Cesàro function space

$$\mathrm{Ces}(p) = \left\{f : \int_0^\infty \left(\frac{1}{x}\int_0^x |f(t)|dt\right)^p dx < \infty\right\}$$

and the Copson function space

$$\mathrm{Cop}(p) = \left\{f : \int_0^\infty \left(\int_x^\infty \frac{|f(t)|}{t}dt\right)^p dx < \infty\right\}$$

coincide for $p > 1$. He also derives estimates for the norms of the corresponding inclusion operators. The same result, with different estimates, is due to Boas

[19], who in fact obtained the integral analogue of the Askey-Boas theorem mentioned in Section 13. We shall here generalise (the qualitative content of) Boas' result using the blocking technique of Section 15.

For this purpose we consider an arbitrary positive monotonic function s on (a,b) with $\inf_{x\in(a,b)} s(x) = 0$ and $\sup_{x\in(a,b)} s(x) = \infty$.

Theorem 18.1 *Let $0 < p, q < \infty$, and let α, β, α' and β' be real numbers such that $\alpha < -p/q$ and $\alpha' > -p/q$ if s is non-decreasing, $\alpha > -p/q$ and $\alpha' < -p/q$ if s is non-increasing. If $\alpha + \beta = \alpha' + \beta'$, then for any measurable function f on (a,b) the condition*

$$\int_a^b \left(s(x)^\alpha \int_a^x s(t)^\beta |f(t)|^p \, dt \right)^{q/p} |ds(x)| < \infty$$

is equivalent to

$$\int_a^b \left(s(x)^{\alpha'} \int_x^b s(t)^{\beta'} |f(t)|^p \, dt \right)^{q/p} |ds(x)| < \infty.$$

The proof is similar to that of Theorem 13.1 using Theorem 15.1.

The qualitative version of Boas' (and hence of Bennett's) result is recovered when considering $p = 1$ and $s(x) = x$ on $(0, \infty)$.

The dual of the space L_0^p

A measurable function f on an interval I is said to have a Lebesgue point of order p, $p \geq 1$, at $x_0 \in I$ if $f \in L^p(I)$ and

$$\frac{1}{x} \int_{I \cap (x_0 - x, x_0 + x)} |f(t) - f(x_0)|^p \, dt \to 0 \quad \text{as} \quad x \to 0.$$

In 1948, Korenblyum, Kreĭn and Levin [52] introduced the space L_0^p of all measurable functions f on $[0,1]$ with $f(0) = 0$ that have a Lebesgue point of order $p \geq 1$ at 0:

$$L_0^p = \left\{ f \in L^p[0,1] : \frac{1}{x} \int_0^x |f(t)|^p \, dt \to 0 \quad \text{as} \quad x \to 0 \right\}.$$

The space is endowed with its natural norm. Under the canonical bilinear form $\langle f, g \rangle = \int_0^1 f(t) g(t) dt$ the authors determined the dual of L_0^p for $p > 1$ as the space of all measurable functions g on $[0,1]$ such that

$$\|g\| = \int_0^1 \left(-G'(x)\right)^{1/p^*} dx < \infty,$$

where G is the maximal convex function on $(0,1]$ satisfying

$$G(x) \leq \int_x^1 |g(t)|^{p^*} dt \quad \text{for} \quad x \in (0,1],$$

and $\|g\|$ gives the dual norm of g. However, from this representation it is not immediately clear which functions g belong to the dual space. In [94], Tandori showed that for $p \geq 1$ the dual of L_0^p is given canonically as

$$\left\{ g : \sum_{\nu=0}^{\infty} \frac{1}{2^{\nu/p}} \left(\int_{2^{-\nu-1}}^{2^{-\nu}} |g(t)|^{p^*} dt \right)^{1/p^*} < \infty \right\},$$

with the usual interpretation if $p = 1$. Theorem 15.1 now implies the following representation.

Theorem 18.2 *For $1 \leq p < \infty$ the dual of L_0^p under the canonical bilinear form is given by*

$$\left\{ g : \int_0^1 \operatorname*{ess\,sup}_{t \geq x} |g(t)| dx < \infty \right\} \qquad \text{if} \quad p = 1,$$

$$\left\{ g : \int_0^1 \left(\frac{1}{x} \int_x^1 |g(t)|^{p^*} dt \right)^{1/p^*} dx < \infty \right\} \qquad \text{if} \quad p > 1.$$

It is remarkable that in the case $p = 1$ this representation even provides the dual norm, as was shown by Tandori [95]. Of course, the definition of L_0^p also makes sense for $0 < p < 1$, but in that case the dual is trivial, as follows easily from the corresponding result for the spaces L^p.

Spaces very similar to the L_0^p have been introduced by D. Borwein [21] in connection with the strong Cesàro summability of functions. These spaces W_p are the integral analogues of the sequence spaces w_p, see Section 13. Their duals in section form follow from the block form derived in [21] by another application of Theorem 15.1.

Summability of functions at Lebesgue points

Motivated by the problem of recovering a function from its Fourier series or its Fourier transform the convergence of integrals

$$\Phi_n(f,x) = \int_a^b f(t)\varphi_n(x,t)dt$$

as $n \to \infty$ has been widely studied, see [1], [88, Chapter I]. These integrals are sometimes called singular integrals (in the sense of Lebesgue). Many authors have investigated the pointwise convergence of these integrals at Lebesgue points (of order p) of f. It is here that the space L_0^p considered above and its dual come into play. The following is the section version of the corresponding result by Korenblyum, Kreĭn and Levin [52] and Tandori [94].

Theorem 18.3 *Let* $1 < p < \infty$, *and let* (φ_n) *be a sequence of measurable functions on* $[0,1]$. *Then we have*

$$\lim_{n\to\infty} \int_0^1 f(t)\varphi_n(t)dt = f(0)$$

for every function $f \in L^p[0,1]$ *that has a Lebesgue point of order* p *at* 0 *if and only if the following conditions are satisfied:*

$$\text{(18.1i)} \qquad \lim_n \int_0^1 \varphi_n(t)dt = 1,$$

$$\text{(18.1ii)} \qquad \lim_n \int_\delta^1 \varphi_n(t)dt = 0 \qquad \textit{for every } \delta > 0,$$

$$\text{(18.1iii)} \qquad \sup_n \int_0^1 \left(\frac{1}{x}\int_x^1 |\varphi_n(t)|^{p^*}dt\right)^{1/p^*} dx < \infty.$$

If one restricts the functions f to those in L_0^p, that is, to those with $f(0) = 0$, the characterising conditions are (18.1ii) and (18.1iii) (cf. [52, Theorem 5] or the proof of [94, Satz I]). As a consequence we see that for a finite interval (a,b) and a point $x \in (a,b)$ we have

$$\lim_{n\to\infty} \Phi_n(f,x) = f(x)$$

for every function $f \in L^p(a,b)$ that has a Lebesgue point of order p at x if and only if the following conditions hold:

$$\lim_n \int_\alpha^\beta \varphi_n(x,t)dt = 1 \qquad \text{whenever } a \le \alpha < x < \beta \le b,$$

$$\sup_n \int_a^b \psi_n(x,t)dt < \infty,$$

where the functions ψ_n are defined by $\psi_n(x,t) = \left(\frac{1}{x-t}\int_a^t |\varphi_n(x,\tau)|^{p^*} d\tau\right)^{1/p^*}$ for $a < t < x$ and $\psi_n(x,t) = \left(\frac{1}{t-x}\int_t^b |\varphi_n(x,\tau)|^{p^*} d\tau\right)^{1/p^*}$ for $x < t < b$.

The corresponding result for $p = 1$ is due to Faddeev [29], see also [1, 4.4.2].

Appendix

Equivalence of Section Conditions with Block Conditions

19 The Discrete Case

The preceding chapters have demonstrated the usefulness of transforming a norm in section form into one in block form, and vice versa. Totik and Vincze [97] and Leindler [65] were the first to consider a closely related problem. Suppose that we are given two norms, one in section form, one in block form. The task then consists in determining if they are equivalent. In this section we use the results of Chapter I to extend the results of Totik-Vincze and Leindler. The analogues for integrals will be given in the next section.

In order to be able to compare a norm in section form with a norm in block form we first compare two different norms in block form. Thus we consider the conditions

(19.1i)
$$\sum_{\nu} \left[\frac{1}{2^{\nu\alpha}} \left(\sum_{k \in I_\nu} |v_k x_k|^p \right)^{1/p} \right]^q < \infty$$

and

(19.1ii)
$$\sum_{\nu} \left[\frac{1}{2^{\nu\beta}} \left(\sum_{k \in J_\nu} |w_k x_k|^p \right)^{1/p} \right]^q < \infty$$

with $0 < p, q \leq \infty$. Throughout this appendix we shall not write down the modifications that are needed in case of infinite p or q. They are the usual ones.

As for the remaining parameters in these conditions, α and β are real numbers, $\boldsymbol{v} = (v_n)$ and $\boldsymbol{w} = (w_n)$ are positive (weight) sequences, and $I_\nu = [m_\nu, m_{\nu+1})$ and $J_\nu = [n_\nu, n_{\nu+1}), \nu \geq 0$, are intervals defined by index sequences $\boldsymbol{m} = (m_\nu)$ and $\boldsymbol{n} = (n_\nu)$, respectively. Obviously, it would suffice to have a weight in only one of the two conditions. We shall keep them both for matters of symmetry.

Theorem 19.1 *The block conditions* (19.1i) *and* (19.1ii) *are equivalent for every sequence* $\boldsymbol{x} = (x_n)$ *if and only if the following conditions hold:*

(I) *There exists some* $K \geq 1$ *such that*

$$K^{-1}\frac{1}{2^{\nu\alpha}} \leq \frac{1}{2^{\mu\beta}}\frac{w_n}{v_n} \leq K\frac{1}{2^{\nu\alpha}}$$

for all $\nu, \mu \geq 0$ *and all* $n \in I_\nu \cap J_\mu$;

(II) *if* $p \neq q$, *there exists an* $N \in \mathbb{N}$ *such that each* $I_\nu, \nu \geq 0$, *meets at most* N *intervals* J_μ, *and each* $J_\mu, \mu \geq 0$, *meets at most* N *intervals* I_ν.

Proof. The closed graph theorem for the identity map between appropriate sequence spaces shows that the equivalence of the block conditions implies, and is implied by, the existence of a constant $K \geq 1$ such that for all sequences $\boldsymbol{x} = (x_n)$ we have

$$(19.2)\quad K^{-1}\sum_\nu\left[\frac{1}{2^{\nu\alpha}}\left(\sum_{k\in I_\nu}|v_k x_k|^p\right)^{1/p}\right]^q \leq \sum_\nu\left[\frac{1}{2^{\nu\beta}}\left(\sum_{k\in J_\nu}|w_k x_k|^p\right)^{1/p}\right]^q \leq K\sum_\nu\left[\frac{1}{2^{\nu\alpha}}\left(\sum_{k\in I_\nu}|v_k x_k|^p\right)^{1/p}\right]^q.$$

Choosing, in particular, $\boldsymbol{x}$ as the nth unit sequence ($n \in \mathbb{N}$) shows the necessity of (I). Now, if (I) holds, then (19.2) is equivalent to

$$(19.3)\quad K^{-1}\sum_\nu\left(\sum_{k\in I_\nu}|y_k|^p\right)^{q/p} \leq \sum_\nu\left(\sum_{k\in J_\nu}|y_k|^p\right)^{q/p} \leq K\sum_\nu\left(\sum_{k\in I_\nu}|y_k|^p\right)^{q/p}$$

for all sequences $\boldsymbol{y}$, with a suitable constant $K \geq 1$. This in turn is equivalent to (II), since (19.3) always holds if $p = q$, while for $p \neq q$ we use the well-known relationship between the norms of l_n^r for different r ($n \in \mathbb{N}$). □

In principle it is now possible to determine for any two norms in block or in section form if they are equivalent or not. One need only transform any norm in section form into one in block form using Section 3, and then apply the theorem. However, when trying to reformulate condition (II) in terms of the parameters appearing in the section norms we were only able to come up with rather impracticable conditions, which is essentially due to the fact that one cannot exclude that some I_ν (or J_ν) are empty. We can circumvent this problem by imposing certain assumptions on w_n/v_n under which parts or all of condition (II) follow from condition (I).

Remark 19.2 (i) Suppose that $\beta \neq 0$ and that there is some $K \geq 1$ such that

$$K^{-1}\frac{w_{m_\nu}}{v_{m_\nu}} \leq \frac{w_n}{v_n} \leq K\frac{w_{m_\nu}}{v_{m_\nu}} \tag{19.4}$$

holds for all $\nu \geq 0$ and $n \in I_\nu$. Then condition (I) implies the first half of condition (II). By symmetry, under a corresponding assumption, condition (I) implies the second half of condition (II).

In order to prove the claim, suppose that I_ν meets both J_{μ_1} and J_{μ_2}. Then condition (I) and (19.4) imply that for any $n_i \in I_\nu \cap J_{\mu_i} (i = 1, 2)$ we have

$$2^{\mu_2\beta} \leq K\frac{w_{n_2}}{v_{n_2}}2^{\nu\alpha} \leq K\frac{w_{m_\nu}}{v_{m_\nu}}2^{\nu\alpha} \leq K\frac{w_{n_1}}{v_{n_1}}2^{\nu\alpha} \leq K\,2^{\mu_1\beta}$$

and also

$$2^{\mu_1\beta} \leq K\,2^{\mu_2\beta}.$$

Hence, $\mu_2 - \mu_1$ stays bounded with a bound that is independent of ν.

(ii) Suppose that $\beta \neq 0$ and that there are $\gamma \in \mathbb{R}$ and $K \geq 1$ such that

$$K^{-1}\frac{1}{2^{\nu\gamma}} \leq \frac{w_n}{v_n} \leq K\frac{1}{2^{\nu\gamma}} \tag{19.5}$$

holds for all $\nu \geq 0$ and $n \in I_\nu$. Under this stronger assumption, condition (II) can be dropped completely. This is so, in particular, if $\boldsymbol{w} = \boldsymbol{v}$ and $\beta \neq 0$.

The proof is similar to that in (i). We note that one may replace the assumption $\beta \neq 0$ equivalently by $\alpha \neq \gamma$ here.

This remark, together with Theorem 19.1, leads to a characterisation of the equivalence of a condition in block form, that is

$$\sum_\nu \left[\frac{1}{2^{\nu\alpha}}\left(\sum_{k\in I_\nu} |v_k x_k|^p\right)^{1/p}\right]^q < \infty, \tag{19.6i}$$

with a condition in section form, that is

$$\sum_n \left[a_n \left(\sum_{k=1}^{n} |w_k x_k|^p \right)^{1/p} \right]^q < \infty \tag{19.6ii}$$

or

$$\sum_n \left[a_n \left(\sum_{k=n}^{\infty} |w_k x_k|^p \right)^{1/p} \right]^q < \infty, \tag{19.6iii}$$

at least for certain $\boldsymbol{v}$ and $\boldsymbol{w}$. Here, $0 < p, q \leq \infty$, α is a real number, $\boldsymbol{v} = (v_n)$ and $\boldsymbol{w} = (w_n)$ are positive sequences, $\boldsymbol{a} = (a_n)$ is a non-negative sequence, and the $I_\nu = [m_\nu, m_{\nu+1}), \nu \geq 0$, are intervals defined by an index sequence $\boldsymbol{m} = (m_\nu)$.

Theorem 19.3 *Suppose that no $I_\nu, \nu \geq 0$, is empty and that there is some $K \geq 1$ such that*

$$K^{-1} \frac{w_{m_\nu}}{v_{m_\nu}} \leq \frac{w_n}{v_n} \leq K \frac{w_{m_\nu}}{v_{m_\nu}} \tag{19.7}$$

holds for all $\nu \geq 0$ and $n \in I_\nu$.

(a) *The conditions* (19.6i) *and* (19.6ii) *are equivalent for every sequence $\boldsymbol{x} = (x_n)$ if and only if the following conditions are satisfied.*

(I) *There exists some $K \geq 1$ such that*

$$K^{-1} \frac{1}{2^{\nu\alpha}} \leq \frac{w_n}{v_n} \left(\sum_{k=n}^{\infty} a_k^q \right)^{1/q} \leq K \frac{1}{2^{\nu\alpha}}$$

holds for all $\nu \geq 0$ and $n \in I_\nu$;

(II) *if $p \neq q$, there exists an $N \in \mathbb{N}$ such that*

$$\limsup_{\nu \to \infty} \frac{\sum_{k=m_{\nu+N}}^{\infty} a_k^q}{\sum_{k=m_\nu}^{\infty} a_k^q} < 1.$$

(b) *For $\boldsymbol{a} \neq \boldsymbol{0}$, the conditions* (19.6i) *and* (19.6iii) *are equivalent for every sequence $\boldsymbol{x} = (x_n)$ if and only if the following conditions are satisfied.*

(I) *There exists some $K \geq 1$ such that*

$$K^{-1} \frac{1}{2^{\nu\alpha}} \leq \frac{w_n}{v_n} \left(\sum_{k=1}^{n} a_k^q \right)^{1/q} \leq K \frac{1}{2^{\nu\alpha}}$$

holds for all $n \in I_\nu$, where ν is sufficiently large;

(II) *if* $p \neq q$, *there exists an* $N \in \mathbb{N}$ *such that*

$$\limsup_{\nu\to\infty} \frac{\sum_{k=1}^{m_\nu} a_k^q}{\sum_{k=1}^{m_\nu+N} a_k^q} < 1.$$

Proof. (a) If $q \neq \infty$ and $\boldsymbol{a} \in l^q \setminus \varphi$, then the result follows from Theorem 19.1 and Remark 19.2 after transforming (19.6ii) into block form with the aid of Theorem 3.1. We only remark that instead of the condition in (II) one obtains the existence of an $N \in \mathbb{N}$ such that $\sum_{k=m_\nu+N}^{\infty} a_k^q / \sum_{k=m_\nu}^{\infty} a_k^q \leq 1/2$, which amounts to the same.

On the other hand, if $\boldsymbol{a} \notin l^q \setminus \varphi$, then by Proposition 5.1 the condition in section form cannot be equivalent to any condition in block form, while for the same $\boldsymbol{a}$ condition (I) always fails.

The same arguments work for $q = \infty$ if either $\boldsymbol{a} \in c_0 \setminus \varphi$ or $\boldsymbol{a} \notin l^\infty \setminus \varphi$. It can be shown as in the proof of Theorem 19.1 that the result is also true in the remaining case of $\boldsymbol{a} \in l^\infty \setminus c_0$.

The proof of (b) is similar. □

Remark 19.4 (i) The assumption that all I_ν be non-empty is no serious restriction. This can always be achieved by re-defining the m_ν. The assumption (19.7) on w_n/v_n, on the other hand, is more restrictive, and it would be of interest to extend the theorem to cover all weights $\boldsymbol{v}$ and $\boldsymbol{w}$. Even so, Theorem 19.3 is strong enough to imply our main results in Chapter I, namely Theorems 2.1 to 2.7. To be sure, one has to repeat parts of the proofs of these theorems to show that the conditions (I) and (II) are satisfied.

(ii) The initial assumptions on I_ν and w_n/v_n were introduced in order to obtain, with the help of Remark 19.2, practicable conditions (II). They were not needed for the condition (I), so that the theorem is valid without restriction if $p = q$. But in this case the result is rather trivial anyway. On the other hand, if (19.7) is replaced by the stronger assumption (19.5) with $\gamma \neq \alpha$, in particular if $\boldsymbol{w} = \boldsymbol{v}$ and $\alpha \neq 0$, then the conditions (II) can be dropped completely.

(iii) In Leindler's terminology, cf. Section 6, the conditions (II) say that, if $p \neq q$, the sequence $\big((\sum_{k=m_\nu}^{\infty} a_k^q)^{1/q}\big)_\nu$ has to be quasi-geometrically decreasing and the sequence $\big((\sum_{k=1}^{m_\nu} a_k^q)^{1/q}\big)_\nu$ has to be quasi-geometrically increasing, respectively.

(iv) Part (a) of the theorem corresponds to the special case $\alpha_m = 1/2^{m\alpha}$ in Theorem 1 of Leindler [65]. We have remarked earlier that the condition on B_m in that result is not in fact a necessary condition, see Section 4. Since the hypotheses of Theorem 19.3 are satisfied in Leindler's setting, we now see that the appropriate condition is that the sequence $(\sum_{k=m}^{\infty} B_k)_m$ (rather than (B_m) itself) is quasi-geometrically decreasing, at least in the case of $\alpha_m = 1/2^{m\alpha}$.

(v) Part (b) of the theorem extends the special case $\lambda_k = 1/2^{k\alpha}$ in Theorem 1 of Totik and Vincze [97], cf. (ii) above.

20 The Integral Analogue

In this section we study the equivalence of section conditions and block conditions involving integrals rather than series. It turns out that the technical difficulties that we faced in the discrete case are largely absent in this setting, so that our results here are more complete.

As in Chapter IV we consider a fixed interval (a, b) with $-\infty \leq a < b \leq \infty$ throughout this section. We start again by comparing two different norms in block form, that is we consider the block conditions

$$\sum_{\nu \in \mathbb{Z}} \left[\frac{1}{2^{\nu\alpha}} \left(\int_{I_\nu} |w_1(t) f(t)|^p \, dt \right)^{1/p} \right]^q < \infty \tag{20.1i}$$

and

$$\sum_{\nu \in \mathbb{Z}} \left[\frac{1}{2^{\nu\beta}} \left(\int_{J_\nu} |w_2(t) f(t)|^p \, dt \right)^{1/p} \right]^q < \infty, \tag{20.1ii}$$

where $0 < p, q \leq \infty$ (cf. our remark after condition (19.1)), α and β are real numbers, and w_1 and w_2 are positive measurable (weight) functions on (a, b). Further, the $I_\nu, \nu \in \mathbb{Z}$, are intervals with end-points t_ν and $t_{\nu+1}$, where $\boldsymbol{t} = (t_\nu)_{\nu \in \mathbb{Z}}$ is a sequence with $a \leq t_\nu \leq t_{\nu+1} \leq b, \lim_{\nu \to -\infty} t_\nu = a$ and $\lim_{\nu \to \infty} t_\nu = b$; the intervals $J_\nu, \nu \in \mathbb{Z}$, are similarly defined by a sequence $\boldsymbol{\tau} = (\tau_\nu)_{\nu \in \mathbb{Z}}$.

We shall say that two intervals meet if their intersection has positive measure. The proofs of the following results are similar to those of Theorem 19.1 and Remark 19.2.

Theorem 20.1 *The block conditions* (20.1i) *and* (20.1ii) *are equivalent for every measurable function f on (a, b) if and only if the following conditions hold:*

(I) *There exists some* $K \geq 1$ *such that we have for all* $\nu, \mu \in \mathbb{Z}$

$$K^{-1}\frac{1}{2^{\nu\alpha}} \leq \frac{1}{2^{\mu\beta}}\frac{w_2(t)}{w_1(t)} \leq K\frac{1}{2^{\nu\alpha}} \quad \textit{a.e. in} \quad I_\nu \cap J_\mu;$$

(II) *if* $p \neq q$, *there exists an* $N \in \mathbb{N}$ *such that each* $I_\nu, \nu \in \mathbb{Z}$, *meets at most* N *intervals* J_μ, *and each* $J_\mu, \mu \in \mathbb{Z}$, *meets at most* N *intervals* I_ν.

Remark 20.2 Under certain assumptions on the function w_2/w_1 one can again drop parts of condition (II). Suppose that $\beta \neq 0$ and that there is some $K \geq 1$ such that for all $\nu \in \mathbb{Z}$ there is some $\tau \in I_\nu$ with

$$K^{-1}\frac{w_2(\tau)}{w_1(\tau)} \leq \frac{w_2(t)}{w_1(t)} \leq K\frac{w_2(\tau)}{w_1(\tau)} \quad \text{a.e. in} \quad I_\nu, \tag{20.2}$$

then condition (I) implies the first half of condition (II). Under a corresponding assumption, condition (I) implies the second half of condition (II).

Suppose that $\beta \neq 0$ and that there are $\gamma \in \mathbb{R}, \gamma \neq \alpha$, and $K \geq 1$ such that for all $\nu \in \mathbb{Z}$ we have

$$K^{-1}\frac{1}{2^{\nu\gamma}} \leq \frac{w_2(t)}{w_1(t)} \leq K\frac{1}{2^{\nu\gamma}} \quad \text{a.e. in} \quad I_\nu, \tag{20.3}$$

then condition (II) may be dropped completely. This is so, in particular, if $w_2 = w_1$ with $\alpha, \beta \neq 0$.

Theorem 20.1, together with Theorems 15.3 and 15.4, leads to a characterisation of the equivalence of a condition in section form with one in block form. Thus we shall compare the block condition

$$\sum_{\nu\in\mathbb{Z}}\left[\frac{1}{2^{\nu\alpha}}\left(\int_{I_\nu}|w_1(t)f(t)|^p\,dt\right)^{1/p}\right]^q < \infty \tag{20.4i}$$

with the section conditions

$$\int_a^b\left[u(x)\left(\int_a^x|w_2(t)f(t)|^p\,dt\right)^{1/p}\right]^q dx < \infty \tag{20.4ii}$$

or

$$\int_a^b\left[u(x)\left(\int_x^b|w_2(t)f(t)|^p\,dt\right)^{1/p}\right]^q dx < \infty, \tag{20.4iii}$$

where $0 < p, q \leq \infty$, α is a real number, w_1 and w_2 are positive measurable functions on (a, b), u is a non-negative measurable function on (a, b), and the intervals $I_\nu, \nu \in \mathbb{Z}$, are defined by a sequence $\boldsymbol{t} = (t_\nu)_{\nu \in \mathbb{Z}}$ as above.

In contrast to the discrete case, the problem of empty intervals J_ν (after the transformation from section form into block form) can only occur for $q = \infty$, and then only if u is discontinuous (see Theorems 15.3 and 15.4). Thus, in the following result we need no a priori assumption on w_2/w_1.

Theorem 20.3 *Suppose that $t_\nu < t_{\nu+1}$ for $t_\nu \neq a$ and $t_{\nu+1} \neq b$ and that u is continuous if $q = \infty$.*

(a) *The conditions* (20.4i) *and* (20.4ii) *are equivalent for every measurable function f on (a, b) if and only if the following conditions are satisfied.*

(I) *There exists some $K \geq 1$ such that for all $\nu \in \mathbb{Z}$*

$$K^{-1}\frac{1}{2^{\nu\alpha}} \leq \frac{w_2(t)}{w_1(t)}\left(\int_t^b u(x)^q dx\right)^{1/q} \leq K\frac{1}{2^{\nu\alpha}} \quad \text{a.e. in} \quad I_\nu;$$

(II) *if $p \neq q$, there exist $M > 0, \rho < 1$ and $N \in \mathbb{N}$ such that*

$$\int_{t_\nu}^b u(x)^q dx \leq M \int_{t_{\nu+1}}^b u(x)^q dx \quad \text{for} \quad \nu \in \mathbb{Z}$$

and

$$\int_{t_{\nu+N}}^b u(x)^q dx \leq \rho \int_{t_\nu}^b u(x)^q dx \quad \text{if} \quad a < t_\nu < t_{\nu+N} < b.$$

(b) *The conditions* (20.4i) *and* (20.4iii) *are equivalent for every measurable function f on (a, b) if and only if the following conditions are satisfied.*

(I) *There exists some $K \geq 1$ such that for all $\nu \in \mathbb{Z}$*

$$K^{-1}\frac{1}{2^{\nu\alpha}} \leq \frac{w_2(t)}{w_1(t)}\left(\int_a^t u(x)^q dx\right)^{1/q} \leq K\frac{1}{2^{\nu\alpha}} \quad \text{a.e. in} \quad I_\nu;$$

(II) *if $p \neq q$, there exist $M > 0, \rho < 1$ and $N \in \mathbb{N}$ such that*

$$\int_a^{t_{\nu+1}} u(x)^q dx \leq M \int_a^{t_\nu} u(x)^q dx \quad \text{for} \quad \nu \in \mathbb{Z}$$

and

$$\int_a^{t_\nu} u(x)^q dx \leq \rho \int_a^{t_{\nu+N}} u(x)^q dx \quad \text{if} \quad a < t_\nu < t_{\nu+N} < b.$$

The proof is a variation of that of Theorem 19.3 and is based on Theorems 15.3, 15.4 and 20.1. What is new here are the first halves of the conditions (II). They arise from the first half of condition (II) in Theorem 20.1 when we take account of the fact that each interval J_ν is non-empty whenever $\tau_\nu \neq a$ and $\tau_{\nu+1} \neq b$. It is important to note that for $q = \infty$ the expressions

$$\sup_{x \geq b} u(x) \qquad \text{and} \qquad \sup_{x \leq a} u(x)$$

that may appear in (II) have to be interpreted as

$$\limsup_{x \to b} u(x) \qquad \text{and} \qquad \limsup_{x \to a} u(x),$$

respectively.

Remark 20.4 As a consequence of Remark 20.2 the whole or parts of condition (II) in (a) or (b) can be dropped under certain assumptions on w_2/w_1. Thus, under the assumption (20.2) the first half of condition (II) is a consequence of condition (I), and under the assumption (20.3) with $\gamma \neq \alpha$ the condition (II) may be dropped completely. This is so in particular if $w_2 = w_1$ and $\alpha \neq 0$.

In contrast to the discrete case we can now go a step further and compare two different conditions in section form. Thus in the next result we shall characterise the equivalence of the condition

$$\text{(20.5i)} \qquad \int_a^b \left[u(x) \left(\int_a^x |w_1(t) f(t)|^p \, dt \right)^{1/p} \right]^q dx < \infty$$

with

$$\text{(20.5ii)} \qquad \int_a^b \left[v(x) \left(\int_x^b |w_2(t) f(t)|^p \, dt \right)^{1/p} \right]^q dx < \infty,$$

as well as the equivalence of

$$\text{(20.6i)} \qquad \int_a^b \left[u_1(x) \left(\int_a^x |w_1(t) f(t)|^p \, dt \right)^{1/p} \right]^q dx < \infty$$

with

$$\text{(20.6ii)} \qquad \int_a^b \left[u_2(x) \left(\int_a^x |w_2(t) f(t)|^p \, dt \right)^{1/p} \right]^q dx < \infty,$$

and the equivalence of

$$\text{(20.7i)} \qquad \int_a^b \left[v_1(x) \left(\int_x^b |w_1(t) f(t)|^p \, dt \right)^{1/p} \right]^q dx < \infty$$

with

$$\text{(20.7ii)} \qquad \int_a^b \left[v_2(x) \left(\int_x^b |w_2(t) f(t)|^p \, dt \right)^{1/p} \right]^q dx < \infty,$$

where $0 < p, q \leq \infty$, w_1 and w_2 are positive measurable functions on (a, b) and u, u_1, u_2, v, v_1 and v_2 are non-negative measurable functions on (a, b). The following result then is a direct consequence of Theorems 15.3, 15.4 and 20.1.

Theorem 20.5 *Assume that* $0 < \int_t^b u(x)^q dx < \infty$ *and* $0 < \int_a^t v(x)^q dx < \infty$ *for all* $t \in (a, b)$ *and that the same conditions hold for* u_1, u_2 *and* v_1, v_2*, respectively. Further assume that all these functions are continuous if* $q = \infty$.

(a) *The conditions* (20.5i) *and* (20.5ii) *are equivalent for every measurable function* f *on* (a, b) *if and only if the following conditions are satisfied.*

(I) *There exists some* $K \geq 1$ *such that for almost all* $t \in (a, b)$ *we have*

$$K^{-1} \int_t^b u(x)^q dx \leq \frac{w_2(t)^q}{w_1(t)^q} \int_a^t v(x)^q dx \leq K \int_t^b u(x)^q dx;$$

(II) *if* $p \neq q$*, there exist* $\varepsilon > 0$ *and* $\rho < 1$ *such that for* $a < t < t + \tau < b$ *we have that*

$$\frac{\int_a^t v(x)^q dx}{\int_a^{t+\tau} v(x)^q dx} \leq \varepsilon \qquad \textit{implies} \qquad \frac{\int_{t+\tau}^b u(x)^q dx}{\int_t^b u(x)^q dx} \leq \rho$$

and

$$\frac{\int_{t+\tau}^b u(x)^q dx}{\int_t^b u(x)^q dx} \leq \varepsilon \qquad \textit{implies} \qquad \frac{\int_a^t v(x)^q dx}{\int_a^{t+\tau} v(x)^q dx} \leq \rho.$$

(b) *The conditions* (20.6i) *and* (20.6ii) *are equivalent for every measurable function* f *on* (a, b) *if and only if the following conditions are satisfied.*

(I) *There exists some* $K \geq 1$ *such that for almost all* $t \in (a, b)$ *we have*

$$K^{-1} \int_t^b u_1(x)^q dx \leq \frac{w_2(t)^q}{w_1(t)^q} \int_t^b u_2(x)^q dx \leq K \int_t^b u_1(x)^q dx;$$

(II) *if* $p \neq q$, *there exist* $\varepsilon > 0$ *and* $\rho < 1$ *such that for* $a < t < t + \tau < b$ *we have that*

$$\frac{\int_{t+\tau}^{b} u_2(x)^q dx}{\int_t^b u_2(x)^q dx} \leq \varepsilon \qquad \textit{implies} \qquad \frac{\int_{t+\tau}^{b} u_1(x)^q dx}{\int_t^b u_1(x)^q dx} \leq \rho$$

and

$$\frac{\int_{t+\tau}^{b} u_1(x)^q dx}{\int_t^b u_1(x)^q dx} \leq \varepsilon \qquad \textit{implies} \qquad \frac{\int_{t+\tau}^{b} u_2(x)^q dx}{\int_t^b u_2(x)^q dx} \leq \rho.$$

(c) *The conditions* (20.7i) *and* (20.7ii) *are equivalent for every measurable function* f *on* (a, b) *if and only if the following conditions are satisfied.*

(I) *There exists some* $K \geq 1$ *such that for almost all* $t \in (a, b)$ *we have*

$$K^{-1} \int_a^t v_1(x)^q dx \leq \frac{w_2(t)^q}{w_1(t)^q} \int_a^t v_2(x)^q dx \leq K \int_a^t v_1(x)^q dx;$$

(II) *if* $p \neq q$, *there exist* $\varepsilon > 0$ *and* $\rho < 1$ *such that for* $a < t < t + \tau < b$ *we have that*

$$\frac{\int_a^t v_2(x)^q dx}{\int_a^{t+\tau} v_2(x)^q dx} \leq \varepsilon \qquad \textit{implies} \qquad \frac{\int_a^t v_1(x)^q dx}{\int_a^{t+\tau} v_1(x)^q dx} \leq \rho$$

and

$$\frac{\int_a^t v_1(x)^q dx}{\int_a^{t+\tau} v_1(x)^q dx} \leq \varepsilon \qquad \textit{implies} \qquad \frac{\int_a^t v_2(x)^q dx}{\int_a^{t+\tau} v_2(x)^q dx} \leq \rho.$$

As a simple application of this result one may show again that the Cesàro function space Ces(p) coincides with the Copson function space Cop(p) for $p > 1$, cf. Section 18.

A concluding remark

We want to end these notes by repeating a remark that we made in the Introduction. As we have emphasised throughout, the blocking technique is limited to providing, in a certain sense, qualitative results (and it is very powerful in that respect). But each time a result has been found in this way, the problem arises of uncovering its quantitative content, that is, of finding the exact value of or good estimates for the hidden best-possible constants. To do this, different and finer methods than the blocking technique are called for. In that sense virtually each of the results in these notes poses a new and open problem.

References

[1] G. ALEXITS, *Konvergenzprobleme der Orthogonalreihen* (VEB Deutscher Verlag der Wissenschaften, Berlin, 1960).

[2] K. F. ANDERSEN AND H. P. HEINIG, Weighted norm inequalities for certain integral operators, *SIAM J. Math. Anal.* 14 (1983), 834–844.

[3] J. M. ANDERSON AND D. GIRELA, Inequalities of Littlewood-Paley type, multipliers and radial growth of the derivative of analytic functions, *J. Reine Angew. Math.* 465 (1995), 11–40.

[4] J. M. ANDERSON AND A. L. SHIELDS, Coefficient multipliers of Bloch functions, *Trans. Amer. Math. Soc.* 224 (1976), 255–265.

[5] R. ASKEY AND R. P. BOAS, Some integrability theorems for power series with positive coefficients, *Mathematical essays dedicated to A. J. Macintyre* (Ohio University Press, Athens, Ohio, 1970), 23–32.

[6] B. AUBERTIN AND J. J. F. FOURNIER, Integrability theorems for trigonometric series, *Studia Math.* 107 (1993), 33–59.

[7] W. BALSER, W. B. JURKAT AND A. PEYERIMHOFF, On linear functionals and summability factors for strong summability, *Canad. J. Math.* 30 (1978), 983–996.

[8] W. BEEKMANN, Review of [67], *Math. Rev.* 51 (1975), # 10948.

[9] P. R. BEESACK AND H. P. HEINIG, Hardy's inequalities with indices less than 1, *Proc. Amer. Math. Soc.* 83 (1981), 532–536.

[10] E. S. BELINSKII, E. R. LIFLYAND AND R. M. TRIGUB, The Banach algebra A^* and its properties, *J. Fourier Anal. Appl.* 3 (1997), 103–129.

[11] G. BENNETT, Lower bounds for matrices, *Lin. Algebra Appl.* 82 (1986), 81–98.

[12] G. BENNETT, Some elementary inequalities, *Quart. J. Math. Oxford Ser. (2)* 38 (1987), 401–425.

[13] G. BENNETT, Some elementary inequalities. II, *Quart. J. Math. Oxford Ser. (2)* 39 (1988), 385–400.

[14] G. BENNETT, Some elementary inequalities. III, *Quart. J. Math. Oxford Ser. (2)* 42 (1991), 149–174.

[15] G. Bennett, Factorizing the classical inequalities, *Mem. Amer. Math. Soc.* 120 (1996), *no.* 576.

[16] O. Blasco, Multipliers on weighted Besov spaces of analytic functions, *Banach spaces (Mérida, 1992)* (American Mathematical Society, Providence, RI, 1993), 23–33.

[17] O. Blasco, Multipliers on spaces of analytic functions, *Canad. J. Math.* 47 (1995), 44–64.

[18] R. P. Boas, *Integrability theorems for trigonometric transforms* (Springer, Berlin, 1967).

[19] R. P. Boas, Some integral inequalities related to Hardy's inequality, *J. Anal. Math.* 23 (1970), 53–63.

[20] F. F. Bonsall, Boundedness of Hankel matrices, *J. London Math. Soc.* 29 (1984), 289–300.

[21] D. Borwein, Linear functionals connected with strong Cesàro summability, *J. London Math. Soc.* 40 (1965), 628–634.

[22] M. Sh. Braverman and V. D. Stepanov, On the discrete Hardy inequality, *Bull. London Math. Soc.* 26 (1994), 283–287.

[23] A. Brown, P. R. Halmos and A. L. Shields, Cesàro operators, *Acta Sci. Math. (Szeged)* 26 (1965), 125–137.

[24] M. Buntinas, Products of sequence spaces, *Analysis* 7 (1987), 293–304.

[25] M. Buntinas and N. Tanović-Miller, New integrability and L^1-convergence classes for even trigonometric series. II, *Approximation Theory (Kecskémet, 1990)* (North-Holland, Amsterdam, 1991), 103–125.

[26] M. Buntinas and N. Tanović-Miller, Integrability classes and summability, *Approximation interpolation and summability (Ramat Aviv, 1990/ Ramat Gan, 1990)* (Bar-Ilan University, Ramat Gan, 1991), 75–88.

[27] E. T. Copson, Note on series of positive terms, *J. London Math. Soc.* 2 (1927), 9–12.

[28] E. T. Copson, Note on series of positive terms, *J. London Math. Soc.* 3 (1928), 49–51.

[29] D. K. Faddeev, Sur la représentation des fonctions sommables au moyen d'intégrales singulières, (Russian), *Mat. Sb. N.S.* 1 (43) (1936), 351–368.

[30] H. G. Feichtinger, Banach convolution algebras of functions. II, *Monatsh. Math.* 87 (1979), 181–207.

[31] H. G. Feichtinger, An elementary approach to Wiener's third Tauberian theorem for the Euclidean n-space, *Symposia Mathematica, Vol. XXIX (Cortona, 1984)* (Academic Press, London, 1987), 267–301.

[32] G. A. Fomin, A class of trigonometric series, *Math. Zametki* 23 (1978), 213–222.

[33] J. J. F. Fournier and J. Stewart, Amalgams of L^p and l^q, *Bull. Amer. Math. Soc.* 13 (1985), 1–21.

[34] J. E. Gilbert, Interpolation between weighted L^p-spaces, *Ark. Mat.* 10 (1972), 235–249.

[35] G. H. Hardy, Notes on some points in the integral calculus. LI, *Messenger Math.* 48 (1918/19), 107–112.

[36] G. H. Hardy, Note on a theorem of Hilbert, *Math. Z.* 6 (1920), 314–317.

[37] G. H. Hardy, Notes on some points in the integral calculus. LX, *Messenger Math.* 54 (1924/25), 150–156.

[38] G. H. Hardy, Remarks on three recent notes in the Journal, *J. London Math. Soc.* 3 (1928), 166–169.

[39] G. H. Hardy and J. E. Littlewood, Elementary theorems concerning power series with positive coefficients and moment constants of positive functions, *J. Reine Angew. Math.* 157 (1927), 141–158.

[40] G. H. Hardy, J. E. Littlewood and G. Pólya, *Inequalities* (Cambridge University Press, Cambridge, 1934).

[41] J. H. Hedlund, Multipliers of H^p spaces, *J. Math. Mech.* 18 (1968/69), 1067–1074.

[42] N. Higaki, Remarks on Hardy's convergence theorem, *Tôhoku Math. J.* 41 (1935), 80–90.

[43] F. Holland, Harmonic analysis on amalgams of L^p and l^q, *J. London Math. Soc.* 10 (1975), 295–305.

[44] A. A. Jagers, A note on Cesàro sequence spaces, *Nieuw Arch. Wisk.* 22 (1974), 113–124.

[45] P. D. Johnson Jr. and R. N. Mohapatra, Inequalities involving lower-triangular matrices, *Proc. London Math. Soc.* 41 (1980), 83–137.

[46] P. D. Johnson Jr. and R. N. Mohapatra, On inequalities related to sequence spaces ces[p,q], *General inequalities, 4 (Oberwolfach, 1983)* (Birkhäuser, Basel, 1984), 191–201.

[47] R. Johnson, Lipschitz spaces, Littlewood-Paley spaces, and convoluteurs, *Proc. London Math. Soc.* 29 (1974), 127–141.

[48] I. Jovanović and V. Rakočević, Multipliers of mixed-norm sequence spaces and measures of noncompactness, *Publ. Inst. Math. (Beograd)* 56(70) (1994), 61–68.

[49] N. J. Kalton, N. T. Peck and J. W. Roberts, *An F-space sampler* (Cambridge University Press, Cambridge, 1984).

[50] N. J. Kalton and J. H. Shapiro, Bases and basic sequences in F-spaces, *Studia Math.* 56 (1976), 47–61.

[51] C. N. Kellogg, An extension of the Hausdorff-Young theorem, *Michigan Math. J.* 18 (1971), 121–127.

[52] B. I. Korenblyum, S. G. Kreĭn and B. Ya. Levin, On certain nonlinear questions of the theory of singular integrals, (Russian), *Dokl. Akad. Nauk SSSR (N.S.)* 62 (1948), 17–20.

[53] G. Köthe, *Topological vector spaces. I* (Springer, Berlin, 1969).

[54] B. Kuttner and I. J. Maddox, On strong convergence factors, *Quart. J. Math. Oxford Ser. (2)* 16 (1965), 165–182.

[55] E. Landau, A note on a theorem concerning series of positive terms, *J. London Math. Soc.* 1 (1926), 38–39.

[56] P. Y. Lee, Cesàro sequence spaces, *Math. Chronicle* 13 (1984), 29–45.

[57] G. Leibowitz, Rhaly matrices, *J. Math. Anal. Appl.* 128 (1987), 272–286.

[58] L. Leindler, Generalization of inequalities of Hardy and Littlewood, *Acta Sci. Math. (Szeged)* 31 (1970), 279–285.

[59] L. Leindler, On relations of coefficient-conditions, *Acta Math. Acad. Sci. Hungar.* 39 (1982), 409–420.

[60] L. Leindler, Further sharpening of inequalities of Hardy and Littlewood, *Acta Sci. Math. (Szeged)* 54 (1990), 285–289.

[61] L. Leindler, On the converses of inequalities of Hardy and Littlewood, *Acta Sci. Math. (Szeged)* 58 (1993), 191–196.

[62] L. Leindler, Some inequalities pertaining to Bennett's results, *Acta Sci. Math. (Szeged)* 58 (1993), 261–279.

[63] L. Leindler, Improvements of some theorems of Mulholland concerning Dirichlet series, *Acta Sci. Math. (Szeged)* 58 (1993), 281–297.

[64] L. Leindler, On power series with positive coefficients, *Anal. Math.* 20 (1994), 205–211.

[65] L. Leindler, On equivalence of coefficient conditions with applications, *Acta Sci. Math. (Szeged)* 60 (1995), 495–514.

[66] V. I. Levin and S. B. Stečkin, Inequalities, *Amer. Math. Soc. Transl.* 14 (1960), 1–29.

[67] K.-P. LIM, Matrix transformation in the Cesàro sequence spaces, *Kyungpook Math. J.* 14 (1974), 221–227.

[68] J. LINDENSTRAUSS AND L. TZAFRIRI, *Classical Banach spaces. I* (Springer, Berlin, 1977).

[69] J. E. LITTLEWOOD, On bounded bilinear forms in an infinite number of variables, *Quart. J. Math. (Oxford)* 1 (1930), 164–174.

[70] D. H. LUECKING, Embedding theorems for spaces of analytic functions via Khinchine's inequality, *Michigan Math. J.* 40 (1993), 333–358.

[71] Y. LUH, Die Räume $l(p), l_\infty(p), c(p), c_0(p), w(p), w_0(p)$ und $w_\infty(p)$. Ein Überblick, *Mitt. Math. Sem. Giessen* 180 (1987), 35–57.

[72] W. LUSKY, On generalized Bergman spaces, *Studia Math.* 119 (1996), 77–95.

[73] I. J. MADDOX, Spaces of strongly summable sequences, *Quart. J. Math. (Oxford) Ser. (2)* 18 (1967), 345–355.

[74] M. MATELJEVIĆ AND M. PAVLOVIĆ, L^p-behavior of power series with positive coefficients and Hardy spaces, *Proc. Amer. Math. Soc.* 87 (1983), 309–316.

[75] Y. MEYER, *Wavelets and operators* (Cambridge University Press, Cambridge, 1992).

[76] D. S. MITRINOVIĆ, J. E. PEČARIĆ AND A. M. FINK, *Inequalities involving functions and their integrals and derivatives* (Kluwer Academic Publishers, Dordrecht, 1991).

[77] A. NAKAMURA, Dual spaces and some properties of $l^q(p), 0 < p, q \le \infty$, *Proc. Fac. Sci. Tokai Univ.* 22 (1987), 11–20.

[78] A. NAKAMURA, Conditional bases in the sequence spaces $l^q(p)$, *Proc. Fac. Sci. Tokai Univ.* 23 (1988), 9–20.

[79] P. N. NG AND P. Y. LEE, On the associate spaces of Cesàro sequence spaces, *Nanta Math.* 9 (1976), 168–170.

[80] Y. OKUYAMA AND T. TSUCHIKURA, On the absolute Riesz summability of orthogonal series, *Anal. Math.* 7 (1981), 199–208.

[81] B. OPIC AND A. KUFNER, *Hardy-type inequalities* (Longman, Harlow, 1990).

[82] A. PIETSCH, Eigenvalues of integral operators. I, *Math. Ann.* 247 (1980), 169–178.

[83] H. R. PITT, A note on bilinear forms, *J. London Math. Soc.* 11 (1936), 174–180.

[84] H. C. RHALY, Terraced matrices, *Bull. London Math. Soc.* 21 (1989), 399–406.

[85] W. L. C. SARGENT, On compact matrix transformations between sectionally bounded BK-spaces, *J. London Math. Soc.* 41 (1966), 79–87.

[86] E. SAWYER, Weighted Lebesgue and Lorentz norm inequalities for the Hardy operator, *Trans. Amer. Math. Soc.* 281 (1984), 329–337.

[87] G. SINNAMON AND V. D. STEPANOV, The weighted Hardy inequality: new proofs and the case $p = 1$, *J. London Math. Soc.* 54 (1996), 89–101.

[88] E. M. STEIN AND G. WEISS, *Introduction to Fourier analysis on Euclidean spaces*, (Princeton University Press, Princeton, 1971).

[89] M. STIEGLITZ AND H. TIETZ, Matrixtransformationen von Folgenräumen. Eine Ergebnisübersicht, *Math. Z.* 154 (1977), 1–16.

[90] W. J. STILES, Some properties of $l_p, 0 < p < 1$, *Studia Math.* 42 (1972), 109–119.

[91] P. W. SY, W. ZHANG AND P. Y. LEE, The dual of Cesàro function spaces, *Glas. Mat. Ser. III* 22(42) (1987), 103–112.

[92] R. TABERSKI, A theorem of Toeplitz type for the class of M-summable sequences, *Bull. Acad. Polon. Sci. Sér. Sci. Math. Astronom. Phys.* 8 (1960), 453–458.

[93] T. TAKAHASHI, A note on inequalities, *Tôhoku Math. J.* 41 (1935), 148–150.

[94] K. TANDORI, Über die Konvergenz singulärer Integrale, *Acta Sci. Math. (Szeged)* 15 (1953/54), 223–230.

[95] K. TANDORI, Über einen speziellen Banachschen Raum, *Publ. Math. Debrecen* 3 (1953/54), 263–268.

[96] K. TANDORI, Über die orthogonalen Funktionen. IX (Absolute Summation), *Acta Sci. Math. (Szeged)* 21 (1960), 292–299.

[97] V. TOTIK AND I. VINCZE, On relations of coefficient conditions, *Acta Sci. Math. (Szeged)* 50 (1986), 93–98.

[98] A. WILANSKY, Summability: the inset, replaceable matrices, the basis in summability space, *Duke Math. J.* 19 (1952), 647–660.

[99] A. WILANSKY, *Modern methods in topological vector spaces* (McGraw-Hill, New York, 1978).

[100] A. WILANSKY, *Summability through functional analysis* (North-Holland, Amsterdam, 1984).

[101] WISKUNDIG GENOOTSCHAP, Programma van jaarlijkse prijsvragen, *Nieuw Arch. Wisk.* 19 (1971), 70–76.

Index

Lecture Notes in Mathematics

For information about Vols. 1–1485
please contact your bookseller or Springer-Verlag

Vol. 1486: L. Arnold, H. Crauel, J.-P. Eckmann (Eds.), Lyapunov Exponents. Proceedings, 1990. VIII, 365 pages. 1991.

Vol. 1487: E. Freitag, Singular Modular Forms and Theta Relations. VI, 172 pages. 1991.

Vol. 1488: A. Carboni, M. C. Pedicchio, G. Rosolini (Eds.), Category Theory. Proceedings, 1990. VII, 494 pages. 1991.

Vol. 1489: A. Mielke, Hamiltonian and Lagrangian Flows on Center Manifolds. X, 140 pages. 1991.

Vol. 1490: K. Metsch, Linear Spaces with Few Lines. XIII, 196 pages. 1991.

Vol. 1491: E. Lluis-Puebla, J.-L. Loday, H. Gillet, C. Soulé, V. Snaith, Higher Algebraic K-Theory: an overview. IX, 164 pages. 1992.

Vol. 1492: K. R. Wicks, Fractals and Hyperspaces. VIII, 168 pages. 1991.

Vol. 1493: E. Benoît (Ed.), Dynamic Bifurcations. Proceedings, Luminy 1990. VII, 219 pages. 1991.

Vol. 1494: M.-T. Cheng, X.-W. Zhou, D.-G. Deng (Eds.), Harmonic Analysis. Proceedings, 1988. IX, 226 pages. 1991.

Vol. 1495: J. M. Bony, G. Grubb, L. Hörmander, H. Komatsu, J. Sjöstrand, Microlocal Analysis and Applications. Montecatini Terme, 1989. Editors: L. Cattabriga, L. Rodino. VII, 349 pages. 1991.

Vol. 1496: C. Foias, B. Francis, J. W. Helton, H. Kwakernaak, J. B. Pearson, H_∞-Control Theory. Como, 1990. Editors: E. Mosca, L. Pandolfi. VII, 336 pages. 1991.

Vol. 1497: G. T. Herman, A. K. Louis, F. Natterer (Eds.), Mathematical Methods in Tomography. Proceedings 1990. X, 268 pages. 1991.

Vol. 1498: R. Lang, Spectral Theory of Random Schrödinger Operators. X, 125 pages. 1991.

Vol. 1499: K. Taira, Boundary Value Problems and Markov Processes. IX, 132 pages. 1991.

Vol. 1500: J.-P. Serre, Lie Algebras and Lie Groups. VII, 168 pages. 1992.

Vol. 1501: A. De Masi, E. Presutti, Mathematical Methods for Hydrodynamic Limits. IX, 196 pages. 1991.

Vol. 1502: C. Simpson, Asymptotic Behavior of Monodromy. V, 139 pages. 1991.

Vol. 1503: S. Shokranian, The Selberg-Arthur Trace Formula (Lectures by J. Arthur). VII, 97 pages. 1991.

Vol. 1504: J. Cheeger, M. Gromov, C. Okonek, P. Pansu, Geometric Topology: Recent Developments. Editors: P. de Bartolomeis, F. Tricerri. VII, 197 pages. 1991.

Vol. 1505: K. Kajitani, T. Nishitani, The Hyperbolic Cauchy Problem. VII, 168 pages. 1991.

Vol. 1506: A. Buium, Differential Algebraic Groups of Finite Dimension. XV, 145 pages. 1992.

Vol. 1507: K. Hulek, T. Peternell, M. Schneider, F.-O. Schreyer (Eds.), Complex Algebraic Varieties. Proceedings, 1990. VII, 179 pages. 1992.

Vol. 1508: M. Vuorinen (Ed.), Quasiconformal Space Mappings. A Collection of Surveys 1960-1990. IX, 148 pages. 1992.

Vol. 1509: J. Aguadé, M. Castellet, F. R. Cohen (Eds.), Algebraic Topology - Homotopy and Group Cohomology. Proceedings, 1990. X, 330 pages. 1992.

Vol. 1510: P. P. Kulish (Ed.), Quantum Groups. Proceedings, 1990. XII, 398 pages. 1992.

Vol. 1511: B. S. Yadav, D. Singh (Eds.), Functional Analysis and Operator Theory. Proceedings, 1990. VIII, 223 pages. 1992.

Vol. 1512: L. M. Adleman, M.-D. A. Huang, Primality Testing and Abelian Varieties Over Finite Fields. VII, 142 pages. 1992.

Vol. 1513: L. S. Block, W. A. Coppel, Dynamics in One Dimension. VIII, 249 pages. 1992.

Vol. 1514: U. Krengel, K. Richter, V. Warstat (Eds.), Ergodic Theory and Related Topics III, Proceedings, 1990. VIII, 236 pages. 1992.

Vol. 1515: E. Ballico, F. Catanese, C. Ciliberto (Eds.), Classification of Irregular Varieties. Proceedings, 1990. VII, 149 pages. 1992.

Vol. 1516: R. A. Lorentz, Multivariate Birkhoff Interpolation. IX, 192 pages. 1992.

Vol. 1517: K. Keimel, W. Roth, Ordered Cones and Approximation. VI, 134 pages. 1992.

Vol. 1518: H. Stichtenoth, M. A. Tsfasman (Eds.), Coding Theory and Algebraic Geometry. Proceedings, 1991. VIII, 223 pages. 1992.

Vol. 1519: M. W. Short, The Primitive Soluble Permutation Groups of Degree less than 256. IX, 145 pages. 1992.

Vol. 1520: Yu. G. Borisovich, Yu. E. Gliklikh (Eds.), Global Analysis – Studies and Applications V. VII, 284 pages. 1992.

Vol. 1521: S. Busenberg, B. Forte, H. K. Kuiken, Mathematical Modelling of Industrial Process. Bari, 1990. Editors: V. Capasso, A. Fasano. VII, 162 pages. 1992.

Vol. 1522: J.-M. Delort, F. B. I. Transformation. VII, 101 pages. 1992.

Vol. 1523: W. Xue, Rings with Morita Duality. X, 168 pages. 1992.

Vol. 1524: M. Coste, L. Mahé, M.-F. Roy (Eds.), Real Algebraic Geometry. Proceedings, 1991. VIII, 418 pages. 1992.

Vol. 1525: C. Casacuberta, M. Castellet (Eds.), Mathematical Research Today and Tomorrow. VII, 112 pages. 1992.

Vol. 1526: J. Azéma, P. A. Meyer, M. Yor (Eds.), Séminaire de Probabilités XXVI. X, 633 pages. 1992.

Vol. 1527: M. I. Freidlin, J.-F. Le Gall, Ecole d'Eté de Probabilités de Saint-Flour XX – 1990. Editor: P. L. Hennequin. VIII, 244 pages. 1992.

Vol. 1528: G. Isac, Complementarity Problems. VI, 297 pages. 1992.

Vol. 1529: J. van Neerven, The Adjoint of a Semigroup of Linear Operators. X, 195 pages. 1992.

Vol. 1530: J. G. Heywood, K. Masuda, R. Rautmann, S. A. Solonnikov (Eds.), The Navier-Stokes Equations II – Theory and Numerical Methods. IX, 322 pages. 1992.

Vol. 1531: M. Stoer, Design of Survivable Networks. IV, 206 pages. 1992.

Vol. 1532: J. F. Colombeau, Multiplication of Distributions. X, 184 pages. 1992.

Vol. 1533: P. Jipsen, H. Rose, Varieties of Lattices. X, 162 pages. 1992.

Vol. 1534: C. Greither, Cyclic Galois Extensions of Commutative Rings. X, 145 pages. 1992.

Vol. 1535: A. B. Evans, Orthomorphism Graphs of Groups. VIII, 114 pages. 1992.

Vol. 1536: M. K. Kwong, A. Zettl, Norm Inequalities for Derivatives and Differences. VII, 150 pages. 1992.

Vol. 1537: P. Fitzpatrick, M. Martelli, J. Mawhin, R. Nussbaum, Topological Methods for Ordinary Differential Equations. Montecatini Terme, 1991. Editors: M. Furi, P. Zecca. VII, 218 pages. 1993.

Vol. 1538: P.-A. Meyer, Quantum Probability for Probabilists. X, 287 pages. 1993.

Vol. 1539: M. Coornaert, A. Papadopoulos, Symbolic Dynamics and Hyperbolic Groups. VIII, 138 pages. 1993.

Vol. 1540: H. Komatsu (Ed.), Functional Analysis and Related Topics, 1991. Proceedings. XXI, 413 pages. 1993.

Vol. 1541: D. A. Dawson, B. Maisonneuve, J. Spencer, Ecole d´ Eté de Probabilités de Saint-Flour XXI - 1991. Editor: P. L. Hennequin. VIII, 356 pages. 1993.

Vol. 1542: J.Fröhlich, Th.Kerler, Quantum Groups, Quantum Categories and Quantum Field Theory. VII, 431 pages. 1993.

Vol. 1543: A. L. Dontchev, T. Zolezzi, Well-Posed Optimization Problems. XII, 421 pages. 1993.

Vol. 1544: M.Schürmann, White Noise on Bialgebras. VII, 146 pages. 1993.

Vol. 1545: J. Morgan, K. O'Grady, Differential Topology of Complex Surfaces. VIII, 224 pages. 1993.

Vol. 1546: V. V. Kalashnikov, V. M. Zolotarev (Eds.), Stability Problems for Stochastic Models. Proceedings, 1991. VIII, 229 pages. 1993.

Vol. 1547: P. Harmand, D. Werner, W. Werner, M-ideals in Banach Spaces and Banach Algebras. VIII, 387 pages. 1993.

Vol. 1548: T. Urabe, Dynkin Graphs and Quadrilateral Singularities. VI, 233 pages. 1993.

Vol. 1549: G. Vainikko, Multidimensional Weakly Singular Integral Equations. XI, 159 pages. 1993.

Vol. 1550: A. A. Gonchar, E. B. Saff (Eds.), Methods of Approximation Theory in Complex Analysis and Mathematical Physics IV, 222 pages, 1993.

Vol. 1551: L. Arkeryd, P. L. Lions, P.A. Markowich, S.R. S. Varadhan. Nonequilibrium Problems in Many-Particle Systems. Montecatini, 1992. Editors: C. Cercignani, M. Pulvirenti. VII, 158 pages 1993.

Vol. 1552: J. Hilgert, K.-H. Neeb, Lie Semigroups and their Applications. XII, 315 pages. 1993.

Vol. 1553: J.-L- Colliot-Thélène, J. Kato, P. Vojta. Arithmetic Algebraic Geometry. Trento, 1991. Editor: E. Ballico. VII, 223 pages. 1993.

Vol. 1554: A. K. Lenstra, H. W. Lenstra, Jr. (Eds.), The Development of the Number Field Sieve. VIII, 131 pages. 1993.

Vol. 1555: O. Liess, Conical Refraction and Higher Microlocalization. X, 389 pages. 1993.

Vol. 1556: S. B. Kuksin, Nearly Integrable Infinite-Dimensional Hamiltonian Systems. XXVII, 101 pages. 1993.

Vol. 1557: J. Azéma, P. A. Meyer, M. Yor (Eds.), Séminaire de Probabilités XXVII. VI, 327 pages. 1993.

Vol. 1558: T. J. Bridges, J. E. Furter, Singularity Theory and Equivariant Symplectic Maps. VI, 226 pages. 1993.

Vol. 1559: V. G. Sprindžuk, Classical Diophantine Equations. XII, 228 pages. 1993.

Vol. 1560: T. Bartsch, Topological Methods for Variational Problems with Symmetries. X, 152 pages. 1993.

Vol. 1561: I. S. Molchanov, Limit Theorems for Unions of Random Closed Sets. X, 157 pages. 1993.

Vol. 1562: G. Harder, Eisensteinkohomologie und die Konstruktion gemischter Motive. XX, 184 pages. 1993.

Vol. 1563: E. Fabes, M. Fukushima, L. Gross, C. Kenig, M. Röckner, D. W. Stroock, Dirichlet Forms. Varenna, 1992. Editors: G. Dell'Antonio, U. Mosco. VII, 245 pages. 1993.

Vol. 1564: J. Jorgenson, S. Lang, Basic Analysis of Regularized Series and Products. IX, 122 pages. 1993.

Vol. 1565: L. Boutet de Monvel, C. De Concini, C. Procesi, P. Schapira, M. Vergne. D-modules, Representation Theory, and Quantum Groups. Venezia, 1992. Editors: G. Zampieri, A. D'Agnolo. VII, 217 pages. 1993.

Vol. 1566: B. Edixhoven, J.-H. Evertse (Eds.), Diophantine Approximation and Abelian Varieties. XIII, 127 pages. 1993.

Vol. 1567: R. L. Dobrushin, S. Kusuoka, Statistical Mechanics and Fractals. VII, 98 pages. 1993.

Vol. 1568: F. Weisz, Martingale Hardy Spaces and their Application in Fourier Analysis. VIII, 217 pages. 1994.

Vol. 1569: V. Totik, Weighted Approximation with Varying Weight. VI, 117 pages. 1994.

Vol. 1570: R. deLaubenfels, Existence Families, Functional Calculi and Evolution Equations. XV, 234 pages. 1994.

Vol. 1571: S. Yu. Pilyugin, The Space of Dynamical Systems with the C^0-Topology. X, 188 pages. 1994.

Vol. 1572: L. Göttsche, Hilbert Schemes of Zero-Dimensional Subschemes of Smooth Varieties. IX, 196 pages. 1994.

Vol. 1573: V. P. Havin, N. K. Nikolski (Eds.), Linear and Complex Analysis – Problem Book 3 – Part I. XXII, 489 pages. 1994.

Vol. 1574: V. P. Havin, N. K. Nikolski (Eds.), Linear and Complex Analysis – Problem Book 3 – Part II. XXII, 507 pages. 1994.

Vol. 1575: M. Mitrea, Clifford Wavelets, Singular Integrals, and Hardy Spaces. XI, 116 pages. 1994.

Vol. 1576: K. Kitahara, Spaces of Approximating Functions with Haar-Like Conditions. X, 110 pages. 1994.

Vol. 1577: N. Obata, White Noise Calculus and Fock Space. X, 183 pages. 1994.

Vol. 1578: J. Bernstein, V. Lunts, Equivariant Sheaves and Functors. V, 139 pages. 1994.

Vol. 1579: N. Kazamaki, Continuous Exponential Martingales and *BMO*. VII, 91 pages. 1994.

Vol. 1580: M. Milman, Extrapolation and Optimal Decompositions with Applications to Analysis. XI, 161 pages. 1994.

Vol. 1581: D. Bakry, R. D. Gill, S. A. Molchanov, Lectures on Probability Theory. Editor: P. Bernard. VIII, 420 pages. 1994.

Vol. 1582: W. Balser, From Divergent Power Series to Analytic Functions. X, 108 pages. 1994.

Vol. 1583: J. Azéma, P. A. Meyer, M. Yor (Eds.), Séminaire de Probabilités XXVIII. VI, 334 pages. 1994.

Vol. 1584: M. Brokate, N. Kenmochi, I. Müller, J. F. Rodriguez, C. Verdi, Phase Transitions and Hysteresis. Montecatini Terme, 1993. Editor: A. Visintin. VII. 291 pages. 1994.

Vol. 1585: G. Frey (Ed.), On Artin's Conjecture for Odd 2-dimensional Representations. VIII, 148 pages. 1994.

Vol. 1586: R. Nillsen, Difference Spaces and Invariant Linear Forms. XII, 186 pages. 1994.

Vol. 1587: N. Xi, Representations of Affine Hecke Algebras. VIII, 137 pages. 1994.

Vol. 1588: C. Scheiderer, Real and Étale Cohomology. XXIV, 273 pages. 1994.

Vol. 1589: J. Bellissard, M. Degli Esposti, G. Forni, S. Graffi, S. Isola, J. N. Mather, Transition to Chaos in Classical and Quantum Mechanics. Montecatini Terme, 1991. Editor: S. Graffi. VII, 192 pages. 1994.

Vol. 1590: P. M. Soardi, Potential Theory on Infinite Networks. VIII, 187 pages. 1994.

Vol. 1591: M. Abate, G. Patrizio, Finsler Metrics – A Global Approach. IX, 180 pages. 1994.

Vol. 1592: K. W. Breitung, Asymptotic Approximations for Probability Integrals. IX, 146 pages. 1994.

Vol. 1593: J. Jorgenson & S. Lang, D. Goldfeld, Explicit Formulas for Regularized Products and Series. VIII, 154 pages. 1994.

Vol. 1594: M. Green, J. Murre, C. Voisin, Algebraic Cycles and Hodge Theory. Torino, 1993. Editors: A. Albano, F. Bardelli. VII, 275 pages. 1994.

Vol. 1595: R.D.M. Accola, Topics in the Theory of Riemann Surfaces. IX, 105 pages. 1994.

Vol. 1596: L. Heindorf, L. B. Shapiro, Nearly Projective Boolean Algebras. X, 202 pages. 1994.

Vol. 1597: B. Herzog, Kodaira-Spencer Maps in Local Algebra. XVII, 176 pages. 1994.

Vol. 1598: J. Berndt, F. Tricerri, L. Vanhecke, Generalized Heisenberg Groups and Damek-Ricci Harmonic Spaces. VIII, 125 pages. 1995.

Vol. 1599: K. Johannson, Topology and Combinatorics of 3-Manifolds. XVIII, 446 pages. 1995.

Vol. 1600: W. Narkiewicz, Polynomial Mappings. VII, 130 pages. 1995.

Vol. 1601: A. Pott, Finite Geometry and Character Theory. VII, 181 pages. 1995.

Vol. 1602: J. Winkelmann, The Classification of Three-dimensional Homogeneous Complex Manifolds. XI, 230 pages. 1995.

Vol. 1603: V. Ene, Real Functions – Current Topics. XIII, 310 pages. 1995.

Vol. 1604: A. Huber, Mixed Motives and their Realization in Derived Categories. XV, 207 pages. 1995.

Vol. 1605: L. B. Wahlbin, Superconvergence in Galerkin Finite Element Methods. XI, 166 pages. 1995.

Vol. 1606: P.-D. Liu, M. Qian, Smooth Ergodic Theory of Random Dynamical Systems. XI, 221 pages. 1995.

Vol. 1607: G. Schwarz, Hodge Decomposition – A Method for Solving Boundary Value Problems. VII, 155 pages. 1995.

Vol. 1608: P. Biane, R. Durrett, Lectures on Probability Theory. Editor: P. Bernard. VII, 210 pages. 1995.

Vol. 1609: L. Arnold, C. Jones, K. Mischaikow, G. Raugel, Dynamical Systems. Montecatini Terme, 1994. Editor: R. Johnson. VIII, 329 pages. 1995.

Vol. 1610: A. S. Üstünel, An Introduction to Analysis on Wiener Space. X, 95 pages. 1995.

Vol. 1611: N. Knarr, Translation Planes. VI, 112 pages. 1995.

Vol. 1612: W. Kühnel, Tight Polyhedral Submanifolds and Tight Triangulations. VII, 122 pages. 1995.

Vol. 1613: J. Azéma, M. Emery, P. A. Meyer, M. Yor (Eds.), Séminaire de Probabilités XXIX. VI, 326 pages. 1995.

Vol. 1614: A. Koshelev, Regularity Problem for Quasilinear Elliptic and Parabolic Systems. XXI, 255 pages. 1995.

Vol. 1615: D. B. Massey, Lê Cycles and Hypersurface Singularities. XI, 131 pages. 1995.

Vol. 1616: I. Moerdijk, Classifying Spaces and Classifying Topoi. VII, 94 pages. 1995.

Vol. 1617: V. Yurinsky, Sums and Gaussian Vectors. XI, 305 pages. 1995.

Vol. 1618: G. Pisier, Similarity Problems and Completely Bounded Maps. VII, 156 pages. 1996.

Vol. 1619: E. Landvogt, A Compactification of the Bruhat-Tits Building. VII, 152 pages. 1996.

Vol. 1620: R. Donagi, B. Dubrovin, E. Frenkel, E. Previato, Integrable Systems and Quantum Groups. Montecatini Terme, 1993. Editors:M. Francaviglia, S. Greco. VIII, 488 pages. 1996.

Vol. 1621: H. Bass, M. V. Otero-Espinar, D. N. Rockmore, C. P. L. Tresser, Cyclic Renormalization and Auto-morphism Groups of Rooted Trees. XXI, 136 pages. 1996.

Vol. 1622: E. D. Farjoun, Cellular Spaces, Null Spaces and Homotopy Localization. XIV, 199 pages. 1996.

Vol. 1623: H.P. Yap, Total Colourings of Graphs. VIII, 131 pages. 1996.

Vol. 1624: V. Brînzănescu, Holomorphic Vector Bundles over Compact Complex Surfaces. X, 170 pages. 1996.

Vol.1625: S. Lang, Topics in Cohomology of Groups. VII, 226 pages. 1996.

Vol. 1626: J. Azéma, M. Emery, M. Yor (Eds.), Séminaire de Probabilités XXX. VIII, 382 pages. 1996.

Vol. 1627: C. Graham, Th. G. Kurtz, S. Méléard, Ph. E. Protter, M. Pulvirenti, D. Talay, Probabilistic Models for Nonlinear Partial Differential Equations. Montecatini Terme, 1995. Editors: D. Talay, L. Tubaro. X, 301 pages. 1996.

Vol. 1628: P.-H. Zieschang, An Algebraic Approach to Association Schemes. XII, 189 pages. 1996.

Vol. 1629: J. D. Moore, Lectures on Seiberg-Witten Invariants. VII, 105 pages. 1996.

Vol. 1630: D. Neuenschwander, Probabilities on the Heisenberg Group: Limit Theorems and Brownian Motion. VIII, 139 pages. 1996.

Vol. 1631: K. Nishioka, Mahler Functions and Transcendence.VIII, 185 pages.1996.

Vol. 1632: A. Kushkuley, Z. Balanov, Geometric Methods in Degree Theory for Equivariant Maps. VII, 136 pages. 1996.

Vol.1633: H. Aikawa, M. Essén, Potential Theory – Selected Topics. IX, 200 pages.1996.

Vol. 1634: J. Xu, Flat Covers of Modules. IX, 161 pages. 1996.

Vol. 1635: E. Hebey, Sobolev Spaces on Riemannian Manifolds. X, 116 pages. 1996.

Vol. 1636: M. A. Marshall, Spaces of Orderings and Abstract Real Spectra. VI, 190 pages. 1996.

Vol. 1637: B. Hunt, The Geometry of some special Arithmetic Quotients. XIII, 332 pages. 1996.

Vol. 1638: P. Vanhaecke, Integrable Systems in the realm of Algebraic Geometry. VIII, 218 pages. 1996.

Vol. 1639: K. Dekimpe, Almost-Bieberbach Groups: Affine and Polynomial Structures. X, 259 pages. 1996.

Vol. 1640: G. Boillat, C. M. Dafermos, P. D. Lax, T. P. Liu, Recent Mathematical Methods in Nonlinear Wave Propagation. Montecatini Terme, 1994. Editor: T. Ruggeri. VII, 142 pages. 1996.

Vol. 1641: P. Abramenko, Twin Buildings and Applications to S-Arithmetic Groups. IX, 123 pages. 1996.

Vol. 1642: M. Puschnigg, Asymptotic Cyclic Cohomology. XXII, 138 pages. 1996.

Vol. 1643: J. Richter-Gebert, Realization Spaces of Polytopes. XI, 187 pages. 1996.

Vol. 1644: A. Adler, S. Ramanan, Moduli of Abelian Varieties. VI, 196 pages. 1996.

Vol. 1645: H. W. Broer, G. B. Huitema, M. B. Sevryuk, Quasi-Periodic Motions in Families of Dynamical Systems. XI, 195 pages. 1996.

Vol. 1646: J.-P. Demailly, T. Peternell, G. Tian, A. N. Tyurin, Transcendental Methods in Algebraic Geometry. Cetraro, 1994. Editors: F. Catanese, C. Ciliberto. VII, 257 pages. 1996.

Vol. 1647: D. Dias, P. Le Barz, Configuration Spaces over Hilbert Schemes and Applications. VII. 143 pages. 1996.

Vol. 1648: R. Dobrushin, P. Groeneboom, M. Ledoux, Lectures on Probability Theory and Statistics. Editor: P. Bernard. VIII, 300 pages. 1996.

Vol. 1649: S. Kumar, G. Laumon, U. Stuhler, Vector Bundles on Curves – New Directions. Cetraro, 1995. Editor: M. S. Narasimhan. VII, 193 pages. 1997.

Vol. 1650: J. Wildeshaus, Realizations of Polylogarithms. XI, 343 pages. 1997.

Vol. 1651: M. Drmota, R. F. Tichy, Sequences, Discrepancies and Applications. XIII, 503 pages. 1997.

Vol. 1652: S. Todorcevic, Topics in Topology. VIII, 153 pages. 1997.

Vol. 1653: R. Benedetti, C. Petronio, Branched Standard Spines of 3-manifolds. VIII, 132 pages. 1997.

Vol. 1654: R. W. Ghrist, P. J. Holmes, M. C. Sullivan, Knots and Links in Three-Dimensional Flows. X, 208 pages. 1997.

Vol. 1655: J. Azéma, M. Emery, M. Yor (Eds.), Séminaire de Probabilités XXXI. VIII, 329 pages. 1997.

Vol. 1656: B. Biais, T. Björk, J. Cvitanić, N. El Karoui, E. Jouini, J. C. Rochet, Financial Mathematics. Bressanone, 1996. Editor: W. J. Runggaldier. VII, 316 pages. 1997.

Vol. 1657: H. Reimann, The semi-simple zeta function of quaternionic Shimura varieties. IX, 143 pages. 1997.

Vol. 1658: A. Pumariño, J. A. Rodríguez, Coexistence and Persistence of Strange Attractors. VIII, 195 pages. 1997.

Vol. 1659: V, Kozlov, V. Maz'ya, Theory of a Higher-Order Sturm-Liouville Equation. XI, 140 pages. 1997.

Vol. 1660: M. Bardi, M. G. Crandall, L. C. Evans, H. M. Soner, P. E. Souganidis, Viscosity Solutions and Applications. Montecatini Terme, 1995. Editors: I. Capuzzo Dolcetta, P. L. Lions. IX, 259 pages. 1997.

Vol. 1661: A. Tralle, J. Oprea, Symplectic Manifolds with no Kähler Structure. VIII, 207 pages. 1997.

Vol. 1662: J. W. Rutter, Spaces of Homotopy Self-Equivalences – A Survey. IX, 170 pages. 1997.

Vol. 1663: Y. E. Karpeshina; Perturbation Theory for the Schrödinger Operator with a Periodic Potential. VII, 352 pages. 1997.

Vol. 1664: M. Väth, Ideal Spaces. V, 146 pages. 1997.

Vol. 1665: E. Giné, G. R. Grimmett, L. Saloff-Coste, Lectures on Probability Theory and Statistics 1996. Editor: P. Bernard. X, 424 pages, 1997.

Vol. 1666: M. van der Put, M. F. Singer, Galois Theory of Difference Equations. VII, 179 pages. 1997.

Vol. 1667: J. M. F. Castillo, M. González, Three-space Problems in Banach Space Theory. XII, 267 pages. 1997.

Vol. 1668: D. B. Dix, Large-Time Behavior of Solutions of Linear Dispersive Equations. XIV, 203 pages. 1997.

Vol. 1669: U. Kaiser, Link Theory in Manifolds. XIV, 167 pages. 1997.

Vol. 1670: J. W. Neuberger, Sobolev Gradients and Differential Equations. VIII, 150 pages. 1997.

Vol. 1671: S. Bouc, Green Functors and *G*-sets. VII, 342 pages. 1997.

Vol. 1673: F. D. Grosshans, Algebraic Homogeneous Spaces and Invariant Theory. VI, 148 pages. 1997.

Vol. 1674: G. Klaas, C. R. Leedham-Green, W. Plesken, Linear Pro-*p*-Groups of Finite Width. VIII, 115 pages. 1997.

Vol. 1676: P. Cembranos, J. Mendoza, Banach Spaces of Vector-Valued Functions. VIII, 118 pages. 1997.

Vol. 1677: N. Proskurin, Cubic Metaplectic Forms and Theta Functions. VIII, 196 pages. 1998.

Vol. 1678: O. Krupková, The Geometry of Ordinary Variational Equations. X, 251 pages. 1997.

Vol. 1679: K.-G. Grosse-Erdmann, The Blocking Technique. Weighted Mean Operators and Hardy's Inequality. IX, 114 pages. 1998.

Vol. 1680: K.-Z. Li, F. Oort, Moduli of Supersingular Abdelian Varieties. V, 116 pages. 1998.